Supercavitation

Igor Nesteruk
Editor

Supercavitation

Advances and Perspectives

A collection dedicated to the 70th jubilee of Yu.N. Savchenko

Springer

Editor
Igor Nesteruk
National Academy of Sciences of Ukraine
Institute of Hydromechanics
Department of Free Boundary Flows
Vul. Zheliabova 8/4
03680 Kyiv
Ukraine
inesteruk@yahoo.com

ISBN 978-3-642-23655-6 e-ISBN 978-3-642-23656-3
DOI 10.1007/978-3-642-23656-3
Springer Heidelberg Dordrecht London New York

Library of Congress Control Number: 2011943754

Printed on acid-free paper

Springer is part of Springer Science+Business Media (www.springer.com)

To the Seventieth Jubilee

of correspondent-member of
the National Academy of sciences of Ukraine

Yuriy Savchenko

Address About Yuriy Savchenko

This collection of scientific works is devoted to the seventieth jubilee of the outstanding scientist correspondent-member of the National Academy of sciences of Ukraine Yuriy Savchenko.

He was born in Kiev in 1940 on 26th July. The first labour experience was obtained on a position of an assistant of a river ship captain. In 1964 Yu. Savchenko had finished education at the mechanical-mathematical faculty of the Kyiv's National university by Taras Shevchenko. From this time his life is inseparably linked with the Institute of Hydromechanics of National Academy of sciences of Ukraine. Beginning from the engineer position, he was awarded by Ph.D. degree in 1970 and by Doctor's degree in 1983. From 1988 he is a Head of Department of Free Boundary Flows. In 1998 Yu. Savchenko was elected as a correspondent-member of the National Academy of sciences of Ukraine.

His collaboration with outstanding academician G. Logvinovich, who was Head of the Institute of hydromechanics during period when Yu. Savchenko was a young specialist, had essential influence on formation of his personality and scientific interests. Investigations of flows with free surfaces were one of basic directions in G. Logvinovich's work and then became the main field of scientific researches of Yu. Savchenko. Before G. Logvinovich removal to the Moscow Yu. Savchenko had already been formed as a chief scientist in the field of dynamics of bodies moving in fluid at presence of free surface, and he became at the head of this scientific direction at the Institute.

Now the Yu. Savchenko's scientific works define substantially the world level of understanding the principles of high-speed motions of bodies in fluid with cavities. He is a scientific leader of priority fundamental direction of investigations of the NAS of Ukraine in this field, in which scientists succeeded firstly to exceed the velocity limit 1,000 m/s and to achieve supersonic velocities for the underwater motion.

His activity has an important expressing in sharply defined preference to the experimental investigations, in management of which he showed considerable engineer talents. Yu. Savchenko is one of founders of the experimental base of the Institute of hydromechanics, his creative elaborations were realised in designs of hydrodynamic tunnels and test rigs. According to his projects the Impulse hydrodynamic tunnels were built for Northrop Grumman Corporation (USA) and Shipbuilding Centre in Wuxi (China).

The Yu. Savchenko's contribution in development of new technologies was awarded to the medal "For labour merit" in 1978, he is a laureate of the prize of the Academy of technologic sciences of Ukraine in 1995 in the field of development of new technologies and a laureate of State prize of Ukraine in the field of science and engineering in 2002. The Yu. Savchenko's scientific investigations are represented in about 100 articles, 40 inventions, monographs and more than 200 scientific reports. Yuriy Savchenko has wide domestic and international recognition, he is an initiator of contacts with scientists and institutions of different countries, a member of organizing committees of international conferences and a scientific leader of researches on supercavitation according to contracts with organizations in USA, Germany, China and Singapore.

The activity and life principles of Yu. Savchenko essentially promote the climate of creative search, mutual respect and exactingness formed in the collective. He is a considerable authority in the collective.

Collaborators of the Institute, his colleagues in the research work sincerely congratulate him and wish him happiness and new creative achievements.

Director of the Institute of Hydromechanics of NAS of Ukraine, academician Victor Grinchenko

Contents

Mathematical Theory and Numerical Simulation of Bubbly Cavitating Nozzle Flows

Can F. Delale, Şenay Pasinlioğlu, and Zafer Başkaya

Abstract
Unsteady quasi-one-dimensional and two-dimensional bubbly cavitating nozzle flows are considered using a homogeneous bubbly flow model. For quasi-one-dimensional nozzle flows, the system of model equations is reduced to two evolution equations for the flow speed and bubble radius and the initial and boundary value problems for the evolution equations are formulated. Results obtained for quasi-one-dimensional nozzle flows capture the measured pressure losses due to cavitation, but they turn out to be insufficient in describing the two-dimensional structures. For this reason, model equations for unsteady two-dimensional bubbly cavitating nozzle flows are considered and, by suitable decoupling, they are reduced to evolution equations for the bubble radius and for the velocity field, the latter being determined by an integro-partial differential system for the unsteady acceleration. This integro-partial differential system constitutes the fundamental equations for the evolution of the dilation and vorticity in two-dimensional cavitating nozzle flows. The initial and boundary value problem of the evolution equations are then discussed and a method to integrate the equations is introduced.

1 Introduction

Cavitating flows through converging–diverging nozzles have direct applications in ducts and venturi tubes as well as in Diesel injection nozzles. The first model of bubbly liquid flow through a converging–diverging nozzle was proposed by

C.F. Delale (✉)
Department of Mechanical Engineering, Işık University, Şile, Istanbul, Turkey
e-mail: delale@isikun.edu.tr

I. Nesteruk (ed.), *Supercavitation*,
DOI 10.1007/978-3-642-23656-3_1, © Springer-Verlag Berlin Heidelberg 2012

Tangren et al. [1] using a barotropic model. The problem has been reconsidered by Ishii et al. [2] by taking into account unsteady effects, but still neglecting bubble dynamics. A one-dimensional continuum bubbly flow model that couples spherical bubble dynamics to the flow equations was proposed by van Wijngaarden [3, 4] and was later employed in investigating shock wave structure [5]. Quasi-one-dimensional steady-state solutions of bubbly cavitating flows through converging–diverging nozzles are investigated using the continuum bubbly mixture model [6, 7] by assuming that the gas pressure inside the bubble obeys the polytropic law and by lumping all damping mechanisms by a single damping coefficient in the form of viscous dissipation. These investigations have demonstrated that steady-state solutions are possible only for some range of the cavitation number, with the rest of the parameters kept fixed. Moreover, a recent investigation [8] shows that the temporal stability of these quasi-one-dimensional steady-state solutions suffer from being very sensitive to slight unsteady perturbations. A numerical investigation of unsteady quasi-one-dimensional bubbly cavitating flows have also been carried out [9] showing the possibility of propagating bubbly shock waves in the diverging section of the nozzle.

The aim of this investigation is devoted to a detailed study of unsteady quasi-one-dimensional and two-dimensional bubbly cavitating nozzle flows. For this reason we first discuss the homogeneous bubbly mixture model previously introduced for quasi-one-dimensional steady-state and unsteady cavitating nozzle flows [5–9]. For quasi-one-dimensional cavitating nozzle flows, by a detailed analysis the system of model equations is reduced to two evolution equations for the flow speed and bubble radius and the initial and boundary value problems for the evolution equations are formulated. For this case a numerical algorithm is constructed for the solution of the initial and boundary value problems of evolution equations. Results obtained for quasi-one-dimensional nozzle flows capture the measured pressure losses due to cavitation, but they turn out to be insufficient in describing the two-dimensional structures such as the formation and development of the attached cavity, the formation of the re-entrant jet and bubble cloud shedding and collapse. For this reason model equations for unsteady two-dimensional bubbly cavitating nozzle flows are considered and, by suitable decoupling, they are reduced to evolution equations for the bubble radius and for the velocity field, the latter being determined by an integro-partial differential system for the unsteady acceleration. More importantly, this integro-algebraic partial differential system seems to form the fundamental equations for the evolution of the dilation and vorticity. In particular, the evolution equation of vorticity is shown to be precisely Fridman's equation of vorticity [10], containing terms arising from non-barotropic flow. The initial and boundary value problem of the evolution equations are then discussed and a method to integrate the equations is introduced.

2 Model Equations

2.1 Quasi-One-Dimensional Flows

We consider the unsteady quasi-one-dimensional cavitating nozzle flow of a bubbly mixture and we assume that the initial distributions, inlet conditions and nozzle geometry are such that cavitation can occur in the nozzle. We use a slightly modified version of the homogeneous bubbly mixture model [3–9]. In this model the slip between the bubbles and the liquid as well as the creation (nucleation and bubble fission) and coagulation of bubbles are neglected and spherical bubbles are assumed. These assumptions have been specifically addressed [11–18] and can be taken into account by an improved model. The quasi-one-dimensional unsteady nozzle flow equations then take the form

$$\rho' = \rho_\ell{}'(1-\beta) \tag{1}$$

$$A'\frac{\partial \rho'}{\partial t'} + \frac{\partial}{\partial x'}(\rho' u' A') = 0 \tag{2}$$

$$\rho'\frac{du'}{dt'} = -\frac{\partial p'}{\partial x'} \tag{3}$$

$$\frac{R'^3(1-\beta)}{\beta} = \frac{3}{4\pi\eta_0{}'} = \text{constant.} \tag{4}$$

The above equations are supplemented by a modified Rayleigh-Plesset equation for spherical bubble dynamics, which takes bubble/bubble interactions into account in the mean-field as

$$\begin{aligned} \frac{p'_v - p'}{\rho_\ell{}'} = & \frac{\left[1 + (2/3)\pi\eta'_0(3\Lambda^2 - 1)R'^3\right]}{\left[1 + (4/3)\pi\eta'_0 R'^3\right]} R' \frac{d^2R'}{dt'^2} \\ & + \frac{3}{2}\frac{\left[1 + (8/3)\pi\eta'_0(2\Lambda^2 - 1)R'^3 + (16/9)\pi^2\eta'^2_0\Lambda^2 R'^6\right]}{\left[1 + (4/3)\pi\eta'_0 R'^3\right]^2}\left(\frac{dR'}{dt'}\right)^2 \\ & + \frac{2S'}{\rho'_\ell R'} + \frac{4\mu'_{eff}}{\rho'_\ell R'}\frac{dR'}{dt'} - \frac{p'_{gi}}{\rho'_\ell}\left(\frac{R'_0}{R'}\right)^{3k} \end{aligned} \tag{5}$$

where Λ denotes the bubble/bubble interaction parameter defined by

$$\Lambda = \frac{\Delta r'}{R'} \tag{6}$$

with $\Delta r'$ denoting the radius of influence of interacting bubbles from the center of any fixed bubble [7, 19]. In Eq. 5 a polytropic law for the expansion and compression of the gas inside the gas/vapor bubble is used and all damping mechanisms, in an ad hoc manner, [20–23] are assumed in the form of viscous dissipation, characterized by a single viscosity coefficient μ'_{eff}. Using the normalization

$$\rho = \frac{\rho}{\rho'_\ell} = 1 - \beta, p = \frac{p'}{p'_{i0}}, p_v = \frac{p'_v}{p'_{i0}},$$
$$p_g = \frac{p'_g}{p'_{i0}}, u = \frac{u'}{\sqrt{p'_{i0}/\rho'_\ell}}, \ R = \frac{R'}{R'_{i0}},$$
$$x = \frac{x'}{H'_i}, A = \frac{A'}{A'_i}, \ t = \frac{t'}{\Theta'} = \frac{\sqrt{p'_{i0}/\rho'_\ell}t'}{H'_i}, \tag{7}$$

Eqs. 1–5 take the normalized form

$$\rho = 1 - \beta, \tag{8}$$

$$A\frac{\partial \rho}{\partial t} + \frac{\partial}{\partial x}(\rho u A) = 0, \tag{9}$$

$$\rho\frac{du}{dt} = -\frac{\partial p}{\partial x}, \tag{10}$$

$$R^3\left(\frac{1-\beta}{\beta}\right) = \frac{1-\beta_{i0}}{\beta_{i0}} = \kappa_i^3 \tag{11}$$

and

$$\frac{p_v - p}{L^2} = \frac{\left[1 + (3\Lambda^2 - 1)(R/\kappa_i)^3/2\right]}{\left[1 + (R/\kappa_i)^3\right]} R\frac{d^2R}{dt^2}$$
$$+ \frac{3}{2}\frac{\left[1 + 2(2\Lambda^2 - 1)(R/\kappa_i)^3 + \Lambda^2(R/\kappa_i)^6\right]}{\left[1 + (R/\kappa_i)^3\right]^2}\left(\frac{dR}{dt}\right)^2$$
$$+ \frac{S_0}{L^2R} + \frac{4}{L^2(\mathrm{Re})R}\frac{dR}{dt} - \frac{p_{gi}}{L^2R^{3k}} \tag{12}$$

where L is the ratio of micro scale to macro scale defined by

$$L = \frac{R'_{i0}}{H'_i}, \tag{13}$$

κ_i is a parameter defined in terms of the inlet void fraction β_{i0} by

$$\kappa_i^3 = \frac{1-\beta_{i0}}{\beta_{i0}}, \tag{14}$$

S_0 is the non-dimensional surface tension coefficient defined by

$$S_0 = \frac{2S'}{p'_{i0}R'_{i0}}, \tag{15}$$

and Re is a typical Reynolds number, based on the overall damping coefficient μ'_{eff}, and is defined by

$$\mathrm{Re} = \frac{\rho'_\ell H'_i \sqrt{p'_{i0}/\rho'_\ell}}{\mu'_{eff}}. \tag{16}$$

Furthermore, by eliminating the void fraction β, the mixture density ρ and the mixture pressure p between Eqs. 8–12, we arrive at the evolution equations for the bubble radius $R(x,t)$ and for the flow speed $u(x,t)$ as

$$\frac{\partial R}{\partial t} = -u\frac{\partial R}{\partial x} + \frac{1}{3R^2}(R^3+\kappa_i^3)\left[\left(\frac{1}{A}\frac{dA}{dx}\right)u + \frac{\partial u}{\partial x}\right] \tag{17}$$

and

$$\frac{\partial u}{\partial t} = a(x,t) \tag{18}$$

where the unsteady acceleration satisfies the linear partial differential equation

$$\frac{\partial^2 a}{\partial x^2} + g(R,\frac{\partial R}{\partial x},x)\frac{\partial a}{\partial x} + h(R,\frac{\partial R}{\partial x},x)a = s(R,u,\frac{\partial R}{\partial x},\frac{\partial u}{\partial x},\frac{\partial^2 u}{\partial x^2},\frac{\partial^3 u}{\partial x^3},x) \tag{19}$$

where the functions g, h, and s are given by

$$g(R,\frac{\partial R}{\partial x},x) = \frac{F_1(R)}{F_2(R)}\frac{\partial R}{\partial x} + \frac{1}{A}\frac{dA}{dx}, \tag{20}$$

$$h(R,\frac{\partial R}{\partial x},x) = \frac{F_1(R)}{F_2(R)}\left(\frac{1}{A}\frac{dA}{dx}\right)\frac{\partial R}{\partial x} + \frac{F_3(R)}{F_2(R)} + \frac{d}{dx}\left(\frac{1}{A}\frac{dA}{dx}\right) \tag{21}$$

and

$$
\begin{aligned}
s\left(R,u,\frac{\partial R}{\partial x},\frac{\partial u}{\partial x},\frac{\partial^2 u}{\partial x^2},\frac{\partial^3 u}{\partial x^3},x\right) = -\Bigg\{ & u\frac{\partial^3 u}{\partial x^3} \\
& + \left[\frac{F_1(R)}{F_2(R)}u\frac{\partial R}{\partial x} + \frac{F_4(R)}{F_2(R)}\frac{\partial u}{\partial x} + \frac{F_4(R)}{F_2(R)}u\left(\frac{1}{A}\frac{dA}{dx}\right) + \frac{F_5(R)}{F_2(R)}\right]\frac{\partial^2 u}{\partial x^2} \\
& + \frac{F_6(R)}{F_2(R)}\frac{\partial R}{\partial x}\left(\frac{\partial u}{\partial x}\right)^2 \\
& + \left[\frac{F_7(R)}{F_2(R)}\left(\frac{1}{A}\frac{dA}{dx}\right) - 3\frac{F_3(R)F_5(R)}{RF_2(R)}\right]\frac{\partial R}{\partial x}\frac{\partial u}{\partial x} \\
& + \frac{F_4(R)}{F_2(R)}\left(\frac{1}{A}\frac{dA}{dx}\right)\left(\frac{\partial u}{\partial x}\right)^2 + \left[\frac{F_8(R)}{F_2(R)}u\frac{d}{dx}\left(\frac{1}{A}\frac{dA}{dx}\right)\right. \\
& \left. + \frac{F_9(R)}{F_2(R)}u\left(\frac{1}{A}\frac{dA}{dx}\right)^2 + \frac{F_5(R)}{F_2(R)}\left(\frac{1}{A}\frac{dA}{dx}\right) + \frac{F_3(R)}{F_2(R)}u\right]\frac{\partial u}{\partial x} \\
& + \left[\frac{F_6(R)}{F_2(R)}u^2\left(\frac{1}{A}\frac{dA}{dx}\right)^2 - 3\frac{F_3(R)F_5(R)}{RF_2(R)}u\left(\frac{1}{A}\frac{dA}{dx}\right)\right. \\
& \left. + \frac{F_1(R)}{F_2(R)}u^2\frac{d}{dx}\left(\frac{1}{A}\frac{dA}{dx}\right) + \frac{F_{10}(R)}{F_2(R)}\right]\frac{\partial R}{\partial x} \\
& + \frac{F_9(R)}{F_2(R)}u^2\left(\frac{1}{A}\frac{dA}{dx}\right)\frac{d}{dx}\left(\frac{1}{A}\frac{dA}{dx}\right) + u^2\frac{d^2}{dx^2}\left(\frac{1}{A}\frac{dA}{dx}\right) \\
& + \frac{F_5(R)}{F_2(R)}u\frac{d}{dx}\left(\frac{1}{A}\frac{dA}{dx}\right) + \frac{\partial p_v/\partial x}{F_2(R)}\Bigg\}. \qquad (22)
\end{aligned}
$$

The functions F_j (R); j = 1,2,. . .,10, entering Eqs. 20–22 are given in Appendix A. The solution for the mixture pressure, the void fraction and the density then follow by

$$
\begin{aligned}
p = p_v - \frac{S_0}{R} + \frac{p_{gi}}{R^{3k}} - \frac{L^2\kappa_i^6}{18R^4}\Big[& (6\Lambda^2-1)(R/\kappa_i)^6 \\
& + (6\Lambda^2-2)(R/\kappa_i)^3 - 1\Big]\psi^2 - \frac{4\kappa_i^3}{3(\mathrm{Re})R^3}\left[1+(R/\kappa_i)^3\right]\psi \\
& - \frac{L^2\kappa_i^3}{6R}\left[2+(3\Lambda^2-1)(R/\kappa_i)^3\right]\frac{d\psi}{dt} \qquad (23)
\end{aligned}
$$

and

$$
\beta = 1 - \rho = \frac{R^3}{R^3 + \kappa_i^3} \qquad (24)
$$

where the dilation Ψ is defined by $\Psi = \partial u/\partial x + (1/A\ dA/dx)u$. In particular, Eq. 23 is independent of flow dimensionality and may be helpful for a quantitative comparison of the pressure distributions obtained by different cavitation models, whether they are based on barotropic relations or phase transition models. The steady-state solutions of the model equations are obtained if, in addition to the vanishing of the unsteady acceleration (a=0), $\partial R/\partial t$ also vanishes everywhere for all times. In such a case we precisely recover the steady-state solution [7].

2.2 Two-Dimensional Flows

For the analysis of the 2D (or 3D) structures of partial cavitation and supercavitation observed in experiments, the quasi-one-dimensional model equations discussed above are insufficient. Therefore, the model equations should be extended to multi-dimensional flows. In this section, for simplicity, we introduce the model equations for two-dimensional unsteady bubbly cavitating flows to be able to calculate, at least, some of the 2-D flow structures observed. Using the homogeneous two-phase dispersed flow model and the classical Euler equations, the continuity and momentum equations in two-dimensions take the form

$$\frac{\partial \rho'}{\partial t'} + \frac{\partial}{\partial x'}(\rho' u') + \frac{\partial}{\partial y'}(\rho' v') = 0, \tag{25}$$

$$\rho'\left(\frac{\partial u'}{\partial t'} + u'\frac{\partial u'}{\partial x'} + v'\frac{\partial u'}{\partial y'}\right) = -\frac{\partial p'}{\partial x'} \tag{26}$$

and

$$\rho'\left(\frac{\partial v'}{\partial t'} + u'\frac{\partial v'}{\partial x'} + v'\frac{\partial v'}{\partial y'}\right) = -\frac{\partial p'}{\partial y'} \tag{27}$$

where the mixture density ρ' is given by Eq. 1 and the void fraction β is related to the radius of mono-dispersed spherical bubbles by Eq. 4, assuming there is no bubble creation and coagulation. Equations 25–27 together with Eqs. 1, 4 and the modified Rayleigh–Plesset equation (5) constitute the model equations for unsteady 2-D bubbly cavitating nozzle flows. With the normalization given by Eq. 7 together with $y = y'/H'_i$, the two-dimensional normalized model equations take the form

$$\rho = (1 - \beta), \tag{28}$$

$$\frac{\partial \rho}{\partial t} + \frac{\partial}{\partial x}(\rho u) + \frac{\partial}{\partial y}(\rho v) = 0, \tag{29}$$

$$\rho\left(\frac{\partial u}{\partial t}+u\frac{\partial u}{\partial x}+v\frac{\partial u}{\partial y}\right)=-\frac{\partial p}{\partial x}, \tag{30}$$

$$\rho\left(\frac{\partial v}{\partial t}+u\frac{\partial v}{\partial x}+v\frac{\partial v}{\partial y}\right)=-\frac{\partial p}{\partial y}, \tag{31}$$

and

$$R^3\frac{1-\beta}{\beta}=\frac{1-\beta_{i0}}{\beta_{i0}}=\kappa_i^3. \tag{32}$$

The system of model equations (28)–(32) is completed by the normalized modified Rayleigh–Plessset equation (12). Similar to the procedure above for quasi-one-dimensional flows, we eliminate the normalized mixture density ρ and the void fraction β using the algebraic relations (28) and (32) in the normalized continuity equation (29), and the normalized pressure field between the normalized modified Rayleigh–Plesset equation (12) and the normalized momentum equations (30) and (31). We then arrive at the following system of evolution equations for the normalized radius R and the normalized velocity field (u,v) as

$$\frac{\partial R}{\partial t}=\frac{R^3+\kappa_i^3}{3R^2}\left(\frac{\partial u}{\partial x}+\frac{\partial v}{\partial y}\right)-u\frac{\partial R}{\partial x}-v\frac{\partial R}{\partial y}, \tag{33}$$

$$\frac{\partial u}{\partial t}=a \tag{34}$$

and

$$\frac{\partial v}{\partial t}=b \tag{35}$$

where the unsteady acceleration field (a, b) satisfies the linear system of integro-partial differential equations

$$\frac{\partial a}{\partial x}+\frac{\partial b}{\partial y}-\frac{R\left[2+(3\Lambda^2-1)(R_1/\kappa_i)^3\right]}{R_1\left[2+(3\Lambda^2-1)(R/\kappa_i)^3\right]}\left(\frac{\partial a}{\partial x}+\frac{\partial b}{\partial y}\right)_{y=0}$$
$$-\frac{6R}{L^2\kappa_i^3\left[2+(3\Lambda^2-1)(R/\kappa_i)^3\right]}\int_0^y\frac{b}{\left[1+(R/\kappa_i)^3\right]}dy_1=S_a \tag{36}$$

$$\frac{\partial b}{\partial x}-\frac{\partial a}{\partial y}-\frac{3R^2}{R^3+\kappa_i^3}\left(\frac{\partial R}{\partial x}b-\frac{\partial R}{\partial y}a\right)=S_b \tag{37}$$

where $R_1 = R(x,0,t)$ and the source terms S_a and S_b are given by

$$S_a = -\left(u\frac{\partial\psi}{\partial x} + v\frac{\partial\psi}{\partial y}\right) + \frac{R\left[2 + (3\Lambda^2 - 1)(R_1/\kappa_i)^3\right]}{R_1\left[2 + (3\Lambda^2 - 1)(R/\kappa_i)^3\right]}\left(u\frac{\partial\psi}{\partial x} + v\frac{\partial\psi}{\partial y}\right)_{y=0} + \frac{R}{\left[2 + (3\Lambda^2 - 1)(R/\kappa_i)^3\right]}\int_0^y s_a\, dy_1 \tag{38}$$

and

$$S_b = -\omega\psi - u\frac{\partial\omega}{\partial x} - v\frac{\partial\omega}{\partial y} + \frac{3R^2}{(R^3 + \kappa_i^3)}\left[\left(u\frac{\partial v}{\partial x} + v\frac{\partial v}{\partial y}\right)\frac{\partial R}{\partial x} - \left(u\frac{\partial u}{\partial x} + v\frac{\partial u}{\partial y}\right)\frac{\partial R}{\partial y}\right] \tag{39}$$

where s_a in Eq. 38 is defined by

$$s_a = \frac{6}{L^2\kappa_i^3\left[1 + (R/\kappa_i)^3\right]}\left(u\frac{\partial v}{\partial x} + v\frac{\partial v}{\partial y}\right) + \frac{6}{L^2\kappa_i^3}\frac{\partial p_v}{\partial y} - \left\{\frac{2\kappa_i^3\left[(6\Lambda^2 - 1)(R/\kappa_i)^6 + (6\Lambda^2 - 2)(R/\kappa_i)^3 - 1\right]}{3R^4}\psi + \frac{8\left[1 + (R/\kappa_i)^3\right]}{L^2(Re)R^3}\right\}\frac{\partial\psi}{\partial y} + \left\{\frac{6S_0}{L^2\kappa_i^3R^2} - \frac{18kp_{g0}}{L^2\kappa_i^3R^{3k+1}} + \frac{24}{L^2(Re)R^4}\psi - \frac{2\kappa_i^3\left[(6\Lambda^2 - 1)(R/\kappa_i)^6 - (3\Lambda^2 - 1)(R/\kappa_i)^3 + 2\right]}{3R^5}\psi^2\right\}\frac{\partial R}{\partial y} \tag{40}$$

In Eqs. 38–40, ψ and ω, respectively, denote the dilation (in this case the divergence of the velocity field) and the vorticity and are given by

$$\psi = \frac{\partial u}{\partial x} + \frac{\partial v}{\partial y} \tag{41}$$

and

$$\omega = \frac{\partial v}{\partial x} - \frac{\partial u}{\partial y}. \tag{42}$$

Equations 36 and 37 for the unsteady acceleration field (a, b) constitute the fundamental equations for the transport of the dilation ψ and of the vorticity ω in 2D bubbly cavitating flows. In particular, Eq. 37 is precisely the non-barotropic vorticity transport equation, called the Fridman equation [10], given by

$$\frac{\partial \boldsymbol{\omega}}{\partial t} + (\mathbf{u}.\nabla)\boldsymbol{\omega} = -\psi\boldsymbol{\omega} + (\boldsymbol{\omega}.\nabla)\mathbf{u} + \frac{1}{\rho^2}\nabla\rho \times \nabla p \tag{43}$$

where the term $(\boldsymbol{\omega}.\nabla)\mathbf{u}$ vanishes in 2D. Thus it forms the basis for the generation of vorticity in non-barotropic flows and is responsible for the re-entrant jet in partial cavitation and for all closure models of cavitation. *In the absence of cavitation where the source terms* S_a and S_b *vanish, Eqs.* 36 *and* 37 *reduce to the classical Cauchy-Riemann equations (existence of the complex velocity potential).* The equations for the normalized pressure, normalized density and void fraction then follow from Eqs. 23 and 24 with the dilation now defined by Eq. 41.

3 Initial and Boundary Value Problems for Bubbly Cavitating Nozzle Flows

3.1 Quasi-One-Dimensional Flows

The solution of the hydrodynamic field for unsteady quasi-one-dimensional bubbly cavitating nozzle flows requires the integration of the system of evolution equations (17)–(22) for the bubble radius R and for the flow speed u for a given nozzle geometry (Fig. 1). In this case we first have to specify the initial distributions for the bubble radius and flow speed throughout the whole nozzle, namely

$$R(x,0) = R_0(x) \text{ and } u(x,0) = u_0(x) \text{ for } x_i \le x \le x_e. \tag{44}$$

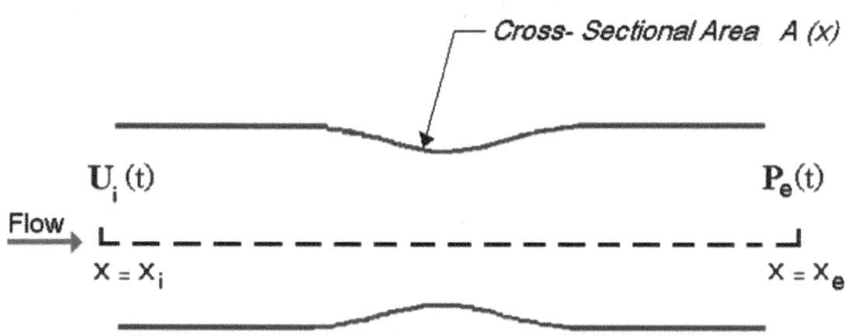

Fig. 1 Geometric configuration of the nozzle employed by Preston et al. [9] and the boundary conditions used for the numerical simulation of quasi-one-dimensional bubbly cavitating nozzle flows

The initial flow field can be taken as the slightly perturbed steady-state quasi-one-dimensional flow field (for the range of parameters where quasi-one-dimensional steady-state solutions are not possible [6, 7], one may start with the incompressible solution supplemented by an everywhere constant bubble radius distribution). To be able to specify the boundary conditions at the nozzle inlet ($x = x_i$) and at the nozzle exit ($x = x_e$), we have to discuss the nature of the evolution equations (17)–(19). In particular, Eq. 17 for the bubble radius evolution is hyperbolic for given flow speed so that we need only to specify the bubble radius at the inlet so that

$$R(x_i, t) = R_i(t) \tag{45}$$

with $R_0(x_i) = R_i(0)$ to avoid a discontinuity in the bubble radius at the nozzle inlet. On the other hand, Eqs. 18 and 19 can be combined into a single evolution equation, coupled to the flow speed and bubble radius, as

$$\begin{aligned}\frac{\partial u}{\partial t} &= a(x,t) = K_1(t)A_1(x,t) + K_2(t)A_2(x,t)\\ &+ A_2(x,t)\int_{x_i}^{x}\frac{[\bar{s}(\xi,t)A_1(\xi,t) + u(\partial^2 u/\partial\xi^2)(\partial A_1/\partial\xi)]}{W(\xi,t)}d\xi\\ &- A_1(x,t)\int_{x_i}^{x}\frac{[\bar{s}(\xi,t)A_2(\xi,t) + u(\partial^2 u/\partial\xi^2)(\partial A_2/\partial\xi)]}{W(\xi,t)}d\xi\end{aligned} \tag{46}$$

where W represents the Wronskian of the two linearly independent solutions A_1 and A_2 of the linear homogeneous equation corresponding to Eq. 19 for the unsteady acceleration a and is given by

$$W(x,t) = A_1\frac{\partial A_2}{\partial x} - A_2\frac{\partial A_1}{\partial x}, \tag{47}$$

and where $K_1(t)$ and $K_2(t)$ are time dependent functions to be determined from the nozzle inlet and exit boundary conditions and $\bar{s}$ is given by

$$\begin{aligned}\bar{s}(x,t) &= \bar{s}(R, u, \frac{\partial R}{\partial x}, \frac{\partial u}{\partial x}, \frac{\partial^2 u}{\partial x^2}, x)\\ &= s(R, u, \frac{\partial R}{\partial x}, \frac{\partial u}{\partial x}, \frac{\partial^2 u}{\partial x^2}, \frac{\partial^3 u}{\partial x^3}, x)\\ &+ u\frac{\partial^3 u}{\partial x^3} + \frac{\partial^2 u}{\partial x^2}[\frac{\partial u}{\partial x} + ug(R, \frac{\partial R}{\partial x}, x)].\end{aligned} \tag{48}$$

In order to evaluate the time dependent functions $K_1(t)$ and $K_2(t)$ in Eq. 46, we consider the appropriate boundary conditions at the inlet and outlet of the nozzle.

For real cavitating flows, either of the following two sets of boundary conditions can be specified:
(a) *The inlet flow speed and exit pressure are specified*, i.e.

$$u(x_i, t) = U_i(t) \text{ and } p(x_e, t) = P_e(t) \text{ for t} \geq 0 \tag{49}$$

together with $U_i(0) = u_0(x_i)$ and $P_e(0) = p(x_e,0)$ to ensure continuity of the solutions.
(b) *The inlet and exit pressures are specified*, i.e.

$$p(x_i, t) = P_i(t) \text{ and } p(x_e, t) = P_e(t) \text{ for t} \geq 0 \tag{50}$$

together with $P_i(0) = p(x_i,0)$ and $P_e(0) = p(x_e,0)$ to ensure continuity of the solutions.

The evaluation of the time dependent functions $K_1(t)$ and $K_2(t)$ in Eq. 46 corresponding to the boundary conditions in each case are given in Appendix B. It should be mentioned that the boundary conditions of case (b) require enormous amount of computation time. Therefore, for simplicity, we adopt the boundary conditions of case (a).

For the numerical method, we first evaluate the unsteady acceleration field by Eq. 46 at every instant t using the flow speed distribution $u(x,t)$ and the radius distribution $R(x,t)$ at that instant, starting with the initial distributions $u_0(x)$ and $R_0(x)$. The homogeneous solutions A_1 and A_2 of Eq. 19 for the unsteady acceleration are obtained by power series methods of second order linear ordinary differential equations with variable coefficients. The time dependent functions $K_1(t)$ and $K_2(t)$ are evaluated using non-reflecting boundary conditions. Using the unsteady acceleration field, the evolution Eq. 18 is integrated using a multi-stage Runge–Kutta method in time to arrive at the flow speed distribution at the next time step. Using the flow speed thus obtained, the first order hyperbolic equation (17) for the bubble radius R is integrated by the classical method of characteristics. Thus the solutions for the flow speed and radius distributions of the evolution equations are obtained for the next time step. The procedure is repeated in a similar manner for all subsequent time steps.

3.2 Two-Dimensional Flows

In order to discuss the solution of the two-dimensional system of evolution equations (33)–(42) of the bubble radius and flow velocity field for cavitating nozzle flows, they should be supplemented by appropriate initial bubble radius and velocity field distributions together with inlet and exit boundary conditions, similar to the case discussed for quasi-one-dimensional flows. In this case the length of the quasi-1D nozzle is elongated in both the inlet and exit directions with corresponding constant inlet and exit areas to ensure uniform inlet and exit boundary conditions across the cross-sectional area at the inlet and exit of the nozzle, as

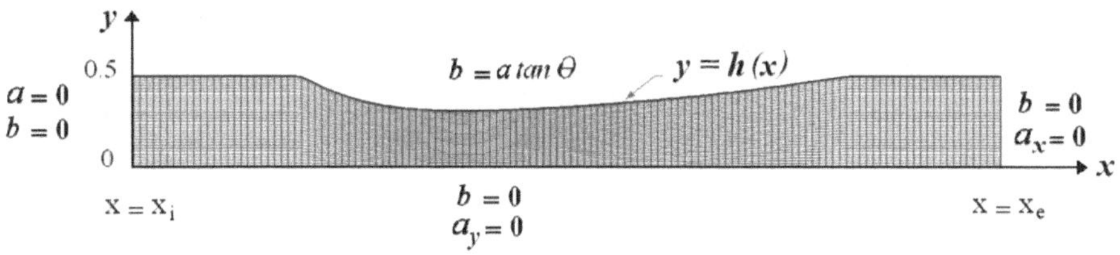

Fig. 2 Nozzle geometry and boundary conditions of the unsteady acceleration field for two-dimensional bubbly cavitating nozzle flows

shown in Fig. 2. We assume a symmetric configuration of the flow field in the y-direction so that it is also sufficient to discuss the solution in the upper symmetric domain of the nozzle. In specifying the initial distributions of the bubble radius and velocity field for the evolution equations, care should be taken to start with irrotational flow in order to access the correct order of magnitude of vorticity generated in the cavitating regime. Therefore, we choose the initial flow field

$$u(x, y, 0) = u_0(x, y) \text{ and } v(x, y, 0) = v_0(x, y) \tag{51}$$

to be irrotational everywhere in the computational domain and uniform and unidirectional (v = 0) at the nozzle inlet and exit (i.e., at $x = x_i$ and $x = x_e$). We also take the initial radius distribution

$$R(x, y, 0) = R_0(x, y) \tag{52}$$

in such a way that it is also uniform at the nozzle inlet and exit. Taking into account the hyperbolicity of Eq. 33 for the bubble radius for given velocity field, we need only to specify the bubble radius at the inlet. Assuming that the inlet bubble radius distribution is uniform in y at all times, we have

$$R(x_i, y, t) = R_i(t) \tag{53}$$

with $R_i(0)$ being equal to the corresponding initial inlet bubble radius to avoid a discontinuity in the bubble radius at the nozzle inlet. Similar to the procedure of quasi-1D flows, we can specify two sets of boundary conditions:

(a) *The inlet flow speed and exit pressure, both uniform, are specified*, i.e.

$$u(x_i, y, t) = U_i(t), \; v(x_i, y, t) = 0 \text{ and } p(x_e, y, t) = P_e(t) \tag{54}$$

for $t \geq 0$ together with $U_i(0)$ and $P_e(0)$ matching the corresponding initial inlet and exit values to ensure continuity of the solutions.

(b) *The uniform inlet and exit pressures are specified*, i.e.

$$p(x_i, y, t) = P_i(t) \text{ and } p(x_e, y, t) = P_e(t) \tag{55}$$

for t $\geq$ 0 together with $P_i(0)$ and $P_e(0)$ matching the corresponding initial inlet and exit values to ensure continuity of the solutions.

The above boundary conditions, similar to the procedure in quasi-one-dimensional flows, should be converted to the boundary conditions for the unsteady acceleration field for the integro-partial differential system, given by Eqs. 36 and 37. For this reason, assuming that the inlet velocity field is uniform and unidirectional and that the bubbles are in mechanical equilibrium at the inlet and exit of the nozzle and using Eq. 23 for the pressure field, we can arrive at the following boundary conditions for the system of Eqs. 36 and 37 in each case:
Case (a) The inlet flow speed and exit pressure, both uniform, are specified.

$$\begin{aligned} &a = 0 \text{ and } b = 0 \text{ at } x = x_i, \ a_x = 0 \text{ and } b = 0 \text{ at } x = x_e, \\ &a_y = 0 \text{ and } b = 0 \text{ at } y = 0, \ b = a \tan\theta \text{ at } \mathrm{y} = \mathrm{h(x)} \end{aligned} \tag{56}$$

where $y = h(x)$ denotes the shape of the upper wall of the nozzle and $tan\theta = dh/dx$. Such a configuration of the boundary conditions are given in Fig. 2.
Case (b) The uniform inlet and exit pressures are specified.

$$\begin{aligned} &a_x = 0 \text{ and } b = 0 \text{ at } x = x_{\mathrm{i}}, \ a_x = 0 \text{ and } b = 0 \text{ at } x = x_{\mathrm{e}}, \\ &a_y = 0 \text{ and } b = 0 \text{ at } y = 0, \ b = a \tan\theta \text{ at } y = h(x). \end{aligned} \tag{57}$$

For the numerical method, similar to the procedure for quasi-1D flows, we first consider the integro-partial differential system of equations, given by Eqs. 36 and 37, subject to boundary conditions given by either Eq. 56 or Eq. 57. The system is solved in two iterative steps. In the first step the integral on the left-hand side of Eq. 36 is set equal to zero and the remaining elliptic system of first order partial differential equations is first discretized by a central finite difference scheme. The resulting linear system of algebraic equations, subject to the boundary conditions given in Eq. 56 or in Eq. 57, are solved by Gauss-Seidel Over Relaxation Method. In the second step, the skipped integral on the left hand side of Eq. 36 is evaluated and treated as a source term. The first step is then repeated to obtain the unsteady acceleration field at that instant. Using a multi-stage Runge–Kutta method in time and the solution for the unsteady acceleration field, the evolution Eqs. 34 and 35 are integrated to yield the velocity field in the next step. Using this velocity field, the hyperbolic evolution equation (33) is integrated by the method of characteristics or by using flux splitting methods to arrive at the bubble radius in the next time step. The numerical scheme is then to be repeated for all subsequent time steps.

4 Results and Discussion

In this section we present results of numerical simulations only for quasi-one-dimensional bubbly cavitating flows. In particular we use two different geometric configurations, whose geometric configurations are shown in Figs. 1, 2, 3,

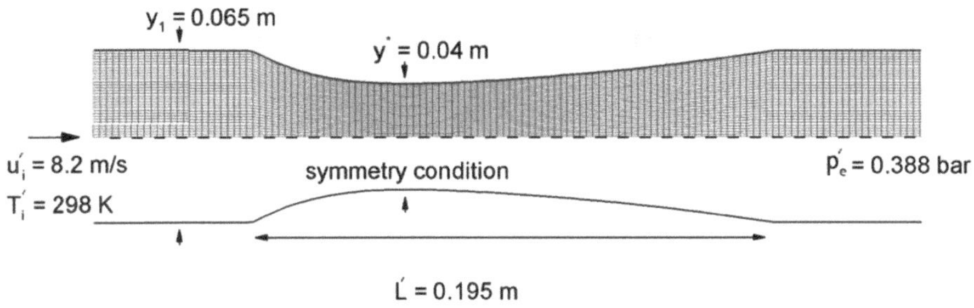

Fig. 3 Geometry and boundary conditions of the nozzle used in the numerical simulation of quasi-one-dimensional flows for comparing the measured wall pressure distributions

respectively. The nozzle employed by Preston et al. [9], whose geometric configuration is shown in Fig. 1, is considered in order to validate our numerical simulation results against their results obtained by some other numerical means. The nozzle whose geometric configuration is shown in Figs. 2 and 3 is employed to compare the pressure distribution obtained for quasi-one-dimensional bubbly cavitating nozzle flows against the measured pressure values at the wall of the nozzle under the same inlet and exit conditions. In both cases we use nozzle inlet velocity and nozzle exit pressure as boundary conditions. We follow the numerical method described in Sect. 3.1 in each case. For the results to be shown we define the cavitation number and the pressure coefficient as

$$\sigma = \frac{p'_{\mathrm{i}0} - p'_{v}}{(1/2)\rho'_{\ell} {u'}^{2}_{i}} \tag{58}$$

and

$$C_p = \frac{p' - p'_{\mathrm{i}0}}{(1/2)\rho'_{\ell} {u'}^{2}_{i}}. \tag{59}$$

To validate the results of our numerical simulation with those of Preston et al. [9], we use the same nozzle employed in their numerical computations with inlet void fraction $\beta_{i0} = 10^{-3}$ and with two different back pressures corresponding to $\sigma = 1.2$ and $\sigma = 0.932$. The results for the pressure coefficients and for the normalized radius are shown in Figs. 4 and 5, respectively. The results for $\sigma = 1.2$ seem to correspond to steady-state conditions whereas those for $\sigma = 0.932$ represent unsteady shocks propagating through the nozzle. The agreement between both numerical predictions is satisfactory.

For the cavitating flow through the nozzle shown in Fig. 3, we consider the two-phase dispersed flow of water with air bubbles with time – averaged inlet flow speed $u'_i = 8.2$ m/s, initial inlet void fraction $\beta_{i0} = 10^{-6}$, initial inlet bubble radius $R'_{i0} =$ 50 μm and time-averaged exit pressure $\mathrm{p}'_{\mathrm{e}} = 0.388$ bar. For the initial field we use a slightly perturbed steady-state distribution for the bubble radius and flow speed.

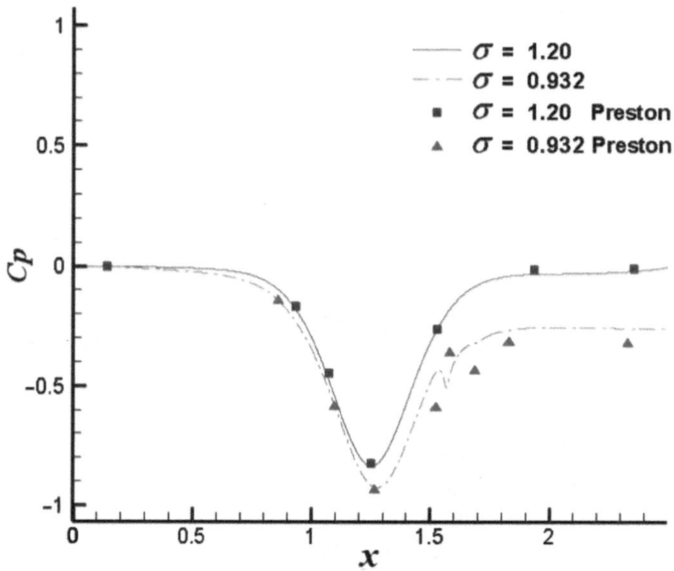

Fig. 4 Comparison of the results for the pressure coefficient obtained by the present numerical simulations against those of Preston et al. [9] for bubbly cavitating flow through the nozzle employed by Preston et al. [9] for two different back pressures corresponding to $\sigma = 1.2$ and $\sigma = 0.932$ with inlet void fraction $\beta_{i0} = 10^{-3}$

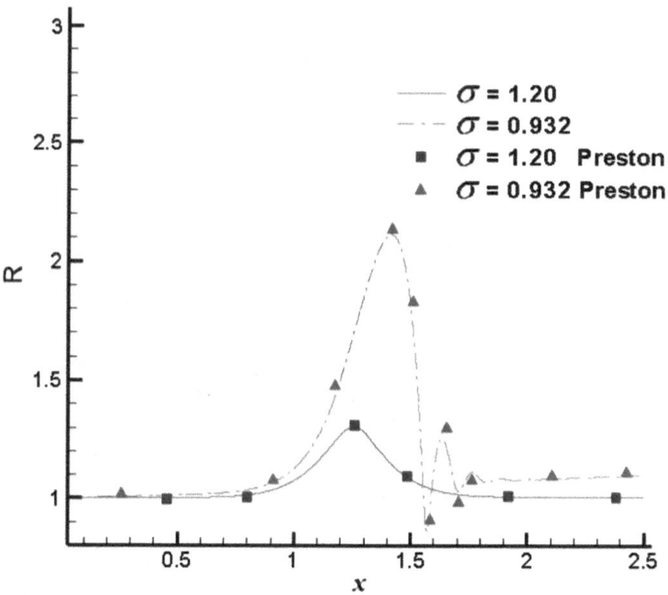

Fig. 5 Comparison of the results for the normalized bubble radius obtained by the present numerical simulations against those of Preston et al. [9] for bubbly cavitating flow through the nozzle employed by Preston et al. [9] for two different back pressures corresponding to $\sigma = 1.2$ and $\sigma = 0.932$ with inlet void fraction $\beta_{i0} = 10^{-3}$

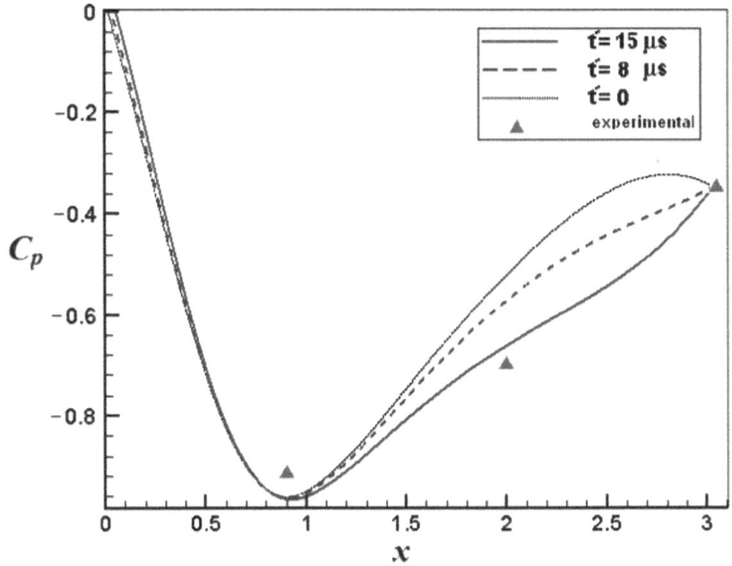

Fig. 6 The unsteady quasi-1D distributions of the pressure coefficient at three instants of time and the time-averaged measured experimental values for the cavitating nozzle flow of water with air bubbles with inlet void fraction $\beta_{i0} = 10^{-6}$, inlet bubble radius $R'_{i0} = 50$ µm, inlet flow speed $u'_i = 8.2$ m/s and exit pressure $p'_e = 0.388$ bar for the nozzle shown in Fig. 3

Under the stated conditions, the steady-state solution shows that the bubbles grow slightly reaching their maximum size and then they return to their initial size. In this case the large growth and violent collapse of the bubbles do not occur and the bubbles seem to be in local mechanical equilibrium [24]. To reach unsteady cavitating flow conditions, we lower the exit pressure until the specified exit pressure under the unsteady cavitating flow conditions is reached. The pressure coefficient, normalized flow speed, normalized bubble radius and normalized unsteady acceleration distributions along the nozzle axis obtained by the bubbly flow model are shown in Figs. 6, 7, 8, 9 at three instants of time at the start of unsteady cavitation. In these figures the transient distributions are ignored and the time $t' = 0$ is artificially set at the begining of unsteady cavitation. It is seen in Fig. 6 that reasonable agreement is achieved between the quasi-one-dimensional unsteady pressure distributions and the measured values from the experiments performed at the Mechanical Engineering Department at Istanbul Technical University under the same conditions. On the other hand, a close examination of the flow speed and radius distributions, shown in Figs. 7 and 8, show that they seem to deviate only slightly fom the initially specified slightly perturbed steady-state distributions, since the cavitation sheets attached to the nozzle walls, in this case, have small thicknesses compared to the nozzle height, thus influencing these distributions only slightly. However, the presence of unsteady cavitation leads to pressure losses which are accommodated by relatively large values of the unsteady acceleration, as shown in Fig. 9. These large values of the unsteady acceleration are balanced by the pressure gradients. They do not contribute to the flow speed significantly because of the very small characteristic times involved.

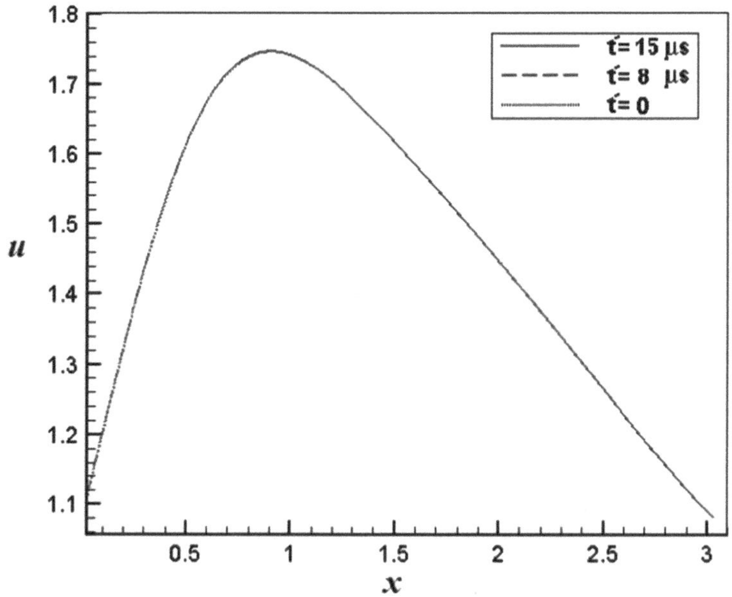

Fig. 7 The unsteady quasi-1D distributions of the normalized flow speed at three instants of time for the cavitating nozzle flow of water with air bubbles with inlet void fraction $\beta_{i0} = 10^{-6}$, inlet bubble radius $R'_{i0} = 50$ μm, inlet flow speed $u'_i = 8.2$ m/s and exit pressure $p'_e = 0.388$ bar for the nozzle shown in Fig. 3

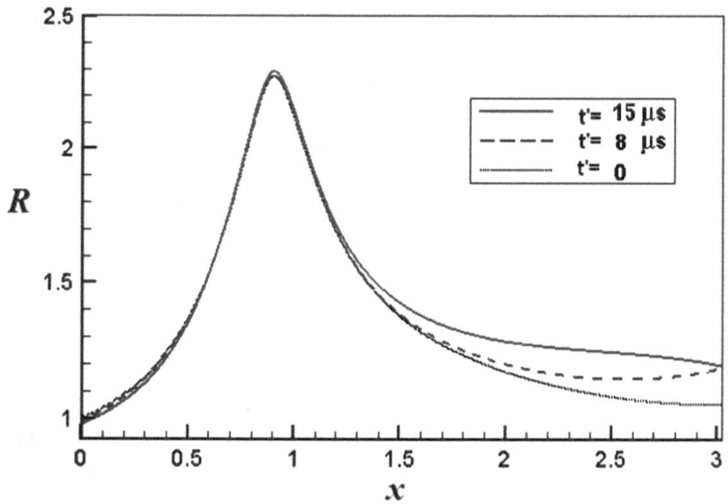

Fig. 8 The unsteady quasi-1D distributions of the normalized bubble radius at three instants of time for the cavitating nozzle flow of water with air bubbles with inlet void fraction $\beta_{i0} = 10^{-6}$, inlet bubble radius $R'_{i0} = 50$ μm, inlet flow speed $u'_i = 8.2$ m/s and exit pressure $p'_e = 0.388$ bar for the nozzle shown in Fig. 3

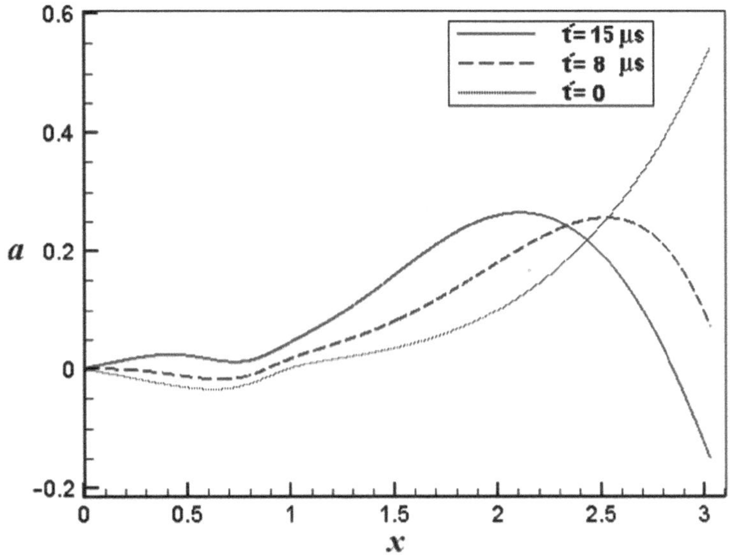

Fig. 9 The unsteady quasi-1D distributions of the normalized unsteady acceleration at three instants of time for the cavitating nozzle flow of water with air bubbles with inlet void fraction $\beta_{i0} = 10^{-6}$, inlet bubble radius $R'_{i0} = 50$ μm, inlet flow speed $u'_i = 8.2$ m/s and exit pressure $p'_e = 0.388$ bar for the nozzle shown in Fig. 3

Conclusions

Model equations for quasi-one-dimensional and two-dimensional bubbly cavitating nozzle flows are presented and the evolution equations for the bubble radius and velocity field in each case are obtained. In particular, in two-dimensional flows the integro-partial differential system of equations for the unsteady acceleration field, which enters the evolution equations for the velocity field, is shown to constitute the fundamental equations of 2D cavitating flows, exhibiting the evolution of the dilation and of the vorticity. The initial/boundary value problems are then formulated for both unsteady quasi-one-dimensional and two-dimensional bubbly cavitating nozzle flows. Results obtained for the unsteady quasi-one-dimensional case show that it is possible to determine the pressure loss due to cavitation in this case.

However, two-dimensional structures of cavitation cannot be determined. Thus the need for a two-dimensional numerical simulation of the model equations is essential. Moreover, the model equations can then be modified to include the boundary layer effect of the flow by using the Navier–Stokes equations for the bubbly mixture and to include bubble nucleation, compressibility and thermal damping effects left out in describing bubble formation and bubble dynamics. These will be the subjects of future investigations.

Acknowledgment This paper is dedicated to Professor Yu.N. Savchenko on the occasion of his 70th birthday.

Appendix A

The functions $F_j\,(R)$; $j = 1,2,\ldots,10$ entering Eqs. 20–22 are

$$F_1(R) = -\frac{L^2\kappa_i^3}{3R^2\left[1+(R/\kappa_i)^3\right]} \times \left[\left(3\Lambda^2-1\right)(R/\kappa_i)^6 + \left(3\Lambda^2-2\right)(R/\kappa_i)^3 - 1\right], \tag{A1}$$

$$F_2(R) = -\frac{L^2\kappa_i^3}{6R}\left[2+\left(3\Lambda^2-1\right)(R/\kappa_i)^3\right], \tag{A2}$$

$$F_3(R) = \frac{1}{\left[1+(R/\kappa_i)^3\right]}, \tag{A3}$$

$$F_4(R) = -\frac{L^2\kappa_i^6}{18R^4}\left[\left(21\Lambda^2-5\right)(R/\kappa_i)^6+\left(12\Lambda^2+2\right)(R/\kappa_i)^3 - 2\right], \tag{A4}$$

$$F_5(R) = -\frac{4\kappa_i^3}{3(\mathrm{Re})R^3}\left[1+(R/\kappa_i)^3\right], \tag{A5}$$

$$F_6(R) = -\frac{L^2\kappa_i^6}{18R^5\left[1+(R/\kappa_i)^3\right]}\left[\left(12\Lambda^2-2\right)(R/\kappa_i)^9 + 6\Lambda^2(R/\kappa_i)^6 - 6\left(\Lambda^2-1\right)(R/\kappa_i)^3 + 4\right], \tag{A6}$$

$$F_7(R) = -\frac{L^2\kappa_i^6}{9R^5\left[1+(R/\kappa_i)^3\right]}\left[\left(21\Lambda^2-5\right)(R/\kappa_i)^9 + \left(15\Lambda^2-6\right)(R/\kappa_i)^6 + \left(-6\Lambda^2+3\right)(R/\kappa_i)^3 + 4\right], \tag{A7}$$

$$F_8(R) = -\frac{L^2\kappa_i^6}{18R^4}\left[\left(39\Lambda^2-11\right)(R/\kappa_i)^6 + \left(12\Lambda^2+14\right)(R/\kappa_i)^3 - 2\right], \tag{A8}$$

$$F_9(R) = -\frac{L^2\kappa_i^6}{18R^4}\left[\left(12\Lambda^2-2\right)(R/\kappa_i)^6 + \left(12\Lambda^2-4\right)(R/\kappa_i)^3 - 2\right], \tag{A9}$$

$$F_{10}(R) = \frac{S_0}{R^2} - \frac{3kP_{g_i}}{R^{3k+1}}. \tag{A10}$$

Appendix B

The boundary conditions presented for the cases (a) and (b) in quasi-one-dimensional flows should be converted to the boundary conditions for the unsteady acceleration field in order to pose a two-point boundary value problem for the unsteady acceleration field given by Eq. 19. For this reason, using Eq. 23 for the pressure distribution in quasi-one-dimensional flows, we can arrive at the boundary conditions for the unsteady acceleration field corresponding to the inlet and exit pressure boundary conditions as

$$\left(\frac{\partial a}{\partial x}\right)_i + \left(\frac{1}{A}\frac{dA}{dx}\right)_i a_i = Q_i(t) \tag{B1}$$

and

$$\left(\frac{\partial a}{\partial x}\right)_e + \left(\frac{1}{A}\frac{dA}{dx}\right)_e a_e = Q_e(t) - \left[\left(\frac{\partial A_1}{\partial x}\right)_e + \left(\frac{1}{A}\frac{dA}{dx}\right)_e (A_1)_e\right] \times \int_{x_i}^{x_e} \frac{\left[\bar{s}(\xi,t)A_2(\xi,t) + u(\partial^2 u/\partial \xi^2)(\partial A_2/\partial \xi)\right]}{W(\xi,t)}\, d\xi \tag{B2}$$

where the functions $Q_i(t)$ and $Q_e(t)$ are defined by

$$\begin{aligned} Q_i(t) = &- U_i\left\{\left(\frac{1}{A}\frac{dA}{dx}\right)_i\left(\frac{\partial u}{\partial x}\right)_i + U_i\left[\frac{d}{dx}\left(\frac{1}{A}\frac{dA}{dx}\right)\right]_i\right\} \\ &- \frac{\kappa_i^3\left[(6\Lambda^2-1)(R_i/\kappa_i)^6 + (6\Lambda^2-2)(R_i/\kappa_i)^3 - 1\right]}{3R_i^3\left[2+(3\Lambda^2-1)(R_i/\kappa_i)^3\right]} \\ &\times \left[\left(\frac{1}{A}\frac{dA}{dx}\right)_i U_i + \left(\frac{\partial u}{\partial x}\right)_i\right]^2 \\ &+ \frac{6R_i}{L^2\kappa_i^3\left[2+(3\Lambda^2-1)(R_i/\kappa_i)^3\right]}\left[(p_v)_i - \frac{S_0}{R_i} + \frac{p_{gi}}{R_i^{3k}} - P_i\right] \\ &- \frac{8\left[1+(R_i/\kappa_i)^3\right]}{L^2(\mathrm{Re})R_i^2\left[2+(3\Lambda^2-1)(R_i/\kappa_i)^3\right]}\left[\left(\frac{1}{A}\frac{dA}{dx}\right)_i U_i + \left(\frac{\partial u}{\partial x}\right)_i\right] \end{aligned} \tag{B3}$$

$$
\begin{aligned}
Q_e(t) = &- U_e\left\{\left(\frac{1}{A}\frac{dA}{dx}\right)_e\left(\frac{\partial u}{\partial x}\right)_e + U_e\left[\frac{d}{dx}\left(\frac{1}{A}\frac{dA}{dx}\right)\right]_e\right\} \\
&- \frac{\kappa_i^3\left[(6\Lambda^2-1)(R_e/\kappa_i)^6 + (6\Lambda^2-2)(R_e/\kappa_i)^3 - 1\right]}{3R_e^3\left[2+(3\Lambda^2-1)(R_e/\kappa_i)^3\right]} \\
&\times\left[\left(\frac{1}{A}\frac{dA}{dx}\right)_e U_e + \left(\frac{\partial u}{\partial x}\right)_e\right]^2 \\
&+ \frac{6R_e}{L^2\kappa_i^3\left[2+(3\Lambda^2-1)(R_e/\kappa_i)^3\right]}\left[(p_v)_e - \frac{S_0}{R_e} + \frac{p_{gi}}{R_e^{3k}} - P_e\right] \\
&- \frac{8\left[1+(R_e/\kappa_i)^3\right]}{L^2(\mathrm{Re})R_e^2\left[2+(3\Lambda^2-1)(R_e/\kappa_i)^3\right]}\left[\left(\frac{1}{A}\frac{dA}{dx}\right)_e U_e + \left(\frac{\partial u}{\partial x}\right)_e\right] \\
&+ \left[\left(\frac{\partial A_1}{\partial x}\right)_e + \left(\frac{1}{A}\frac{dA}{dx}\right)_e (A_1)_e\right] \\
&\times\int_{x_i}^{x_e}\frac{\left[\bar{s}(\xi,t)A_2(\xi,t) + u(\partial^2 u/\partial\xi^2)(\partial A_2/\partial\xi)\right]}{W(\xi,t)}\,d\xi \\
&- \left[\left(\frac{\partial A_2}{\partial x}\right)_e + \left(\frac{1}{A}\frac{dA}{dx}\right)_e (A_2)_e\right] \\
&\times\int_{x_i}^{x_e}\frac{\left[\bar{s}(\xi,t)A_1(\xi,t) + u(\partial^2 u/\partial\xi^2)(\partial A_1/\partial\xi)\right]}{W(\xi,t)}\,d\xi
\end{aligned}
\tag{B4}
$$

The time dependent functions $K_1(t)$ and $K_2(t)$ for case (a) and case (b) boundary conditions in quasi-one-dimensional flows then follow as:
Case (a): The inlet flow speed and exit pressure are specified.

In this case, the functions $K_1(t)$ and $K_2(t)$ satisfy the following equations:

$$
K_1(t)(A_1)_i + K_2(t)(A_2)_i = a_i(t) = \frac{dU_i}{dt} \tag{B5}
$$

$$
K_1(t)\left[\left(\frac{\partial A_1}{\partial x}\right)_e + \left(\frac{1}{A}\frac{dA}{dx}\right)_e (A_1)_e\right] + K_2(t)\left[\left(\frac{\partial A_2}{\partial x}\right)_e + \left(\frac{1}{A}\frac{dA}{dx}\right)_e (A_2)_e\right] = Q_e(t) \tag{B6}
$$

whose solution is given by

$$
K_1(t) = \frac{\left[\left(\frac{\partial A_2}{\partial x}\right)_e + \left(\frac{1}{A}\frac{dA}{dx}\right)_e (A_2)_e\right]\frac{dU_i}{dt} - (A_2)_i Q_e}{\Delta_a(t)} \tag{B7}
$$

and

$$K_2(t) = -\frac{\left[\left(\frac{\partial A_1}{\partial x}\right)_e + \left(\frac{1}{A}\frac{dA}{dx}\right)_e (A_1)_e\right]\frac{dU_i}{dt} - (A_1)_i Q_e}{\Delta_a(t)} \tag{B8}$$

where $\Delta_a(t)$ is given by

$$\begin{aligned}\Delta_a(t) = {} & (A_1)_i\left(\frac{\partial A_2}{\partial x}\right)_e - (A_2)_i\left(\frac{\partial A_1}{\partial x}\right)_e \\ & + \left(\frac{1}{A}\frac{dA}{dx}\right)_e \left[(A_1)_i(A_2)_e - (A_2)_i(A_1)_e\right]\end{aligned} \tag{B9}$$

Case (b): The inlet and exit pressures are specified.

In this case, the functions $K_1(t)$ and $K_2(t)$ satisfy the following equations:

$$\begin{aligned}& K_1(t)\left[\left(\frac{\partial A_1}{\partial x}\right)_i + \left(\frac{1}{A}\frac{dA}{dx}\right)_i (A_1)_i\right] \\ & \quad + K_2(t)\left[\left(\frac{\partial A_2}{\partial x}\right)_i + \left(\frac{1}{A}\frac{dA}{dx}\right)_i (A_2)_i\right] = Q_i(t)\end{aligned} \tag{B10}$$

and

$$\begin{aligned}& K_1(t)\left[\left(\frac{\partial A_1}{\partial x}\right)_e + \left(\frac{1}{A}\frac{dA}{dx}\right)_e (A_1)_e\right] \\ & \quad + K_2(t)\left[\left(\frac{\partial A_2}{\partial x}\right)_e + \left(\frac{1}{A}\frac{dA}{dx}\right)_e (A_2)_e\right] = Q_e(t)\end{aligned} \tag{B11}$$

whose solution is given by

$$\begin{aligned}K_1(t) = & \left\{\left[\left(\frac{\partial A_2}{\partial x}\right)_e + \left(\frac{1}{A}\frac{dA}{dx}\right)_e (A_2)_e\right] Q_i \right. \\ & \left. - \left[\left(\frac{\partial A_2}{\partial x}\right)_i + \left(\frac{1}{A}\frac{dA}{dx}\right)_i (A_2)_i\right] Q_e\right\}[\Delta_b(t)]^{-1}\end{aligned} \tag{B12}$$

and

$$\begin{aligned}K_2(t) = - & \left\{\left[\left(\frac{\partial A_1}{\partial x}\right)_e + \left(\frac{1}{A}\frac{dA}{dx}\right)_e (A_1)_e\right] Q_i \right. \\ & \left. - \left[\left(\frac{\partial A_1}{\partial x}\right)_i + \left(\frac{1}{A}\frac{dA}{dx}\right)_i (A_1)_i\right] Q_e\right\}[\Delta_b(t)]^{-1}\end{aligned} \tag{B13}$$

where $\Delta_b(t)$ is given by

$$\Delta_b(t) = \left[\left(\frac{\partial A_1}{\partial x}\right)_i + \left(\frac{1}{A}\frac{dA}{dx}\right)_i (A_1)_i\right]\left[\left(\frac{\partial A_2}{\partial x}\right)_e + \left(\frac{1}{A}\frac{dA}{dx}\right)_e (A_2)_e\right] - \left[\left(\frac{\partial A_1}{\partial x}\right)_e + \left(\frac{1}{A}\frac{dA}{dx}\right)_e (A_1)_e\right]\left[\left(\frac{\partial A_2}{\partial x}\right)_i + \left(\frac{1}{A}\frac{dA}{dx}\right)_i (A_2)_i\right] \quad \text{(B14)}$$

References

1. Tangren RF, Dodge CH, Seifert HS. Compressibility effects in two-phase flow. J Appl Phys. 1949;20:637–45.
2. Ishii R, Umeda Y, Murata S, Shishido N. Bubbly flows through a converging-diverging nozzle. Phys Fluids A. 1993;5:1630–43.
3. van Wijngaarden L. On the equations of motion for mixtures of liquid and gas bubbles. J Fluid Mech. 1968;33:465–74.
4. van Wijngaarden L. One-dimensional flow of liquids containing small gas bubbles. Ann Rev Fluid Mech. 1972;4:369–96.
5. Noordzij L, van Wijngaarden L. Relaxation effects, caused by the relative motion, on shock waves in gas-bubble/liquid mixtures. J Fluid Mech. 1974;66:115–43.
6. Wang YC, Brennen CE. One dimensional bubbly cavitating flows through a converging-diverging nozzle. ASME J Fluids Eng. 1998;120:166–70.
7. Delale CF, Schnerr GH, Sauer J. Quasi-one-dimensional steady-state cavitating nozzle flows. J Fluid Mech. 2001;427:167–204.
8. Pasinlioğlu Ş, Delale CF, Schnerr GH. On the temporal stability of quasi-one-dimensional steady-state bubbly cavitating nozzle flow solutions. IMA J Appl Math. 2009;74:230–49.
9. Preston AT, Colonius T, Brennen CE. A numerical investigation of unsteady bubbly cavitating nozzle flows. Phys Fluids. 2002;14:300–11.
10. Saffman PG. Vortex dynamics. Cambridge: Cambridge University Press; 1992.
11. Brennen CE. Cavitation and bubble dynamics. Oxford: Oxford University Press; 1995.
12. Wang YC, Chen E. Effect of phase relative motion on critical bubbly flows through a converging-diverging nozzle. Phys Fluids. 2002;14:3215–23.
13. Mørch KA. Cavitation nuclei and bubble formation: a dynamic liquid-solid interface problem. ASME J Fluids Eng. 2000;122:494–8.
14. Delale CF, Hruby J, Marsik F. Homogeneous bubble nucleation in liquids: the classical theory revisited. J Chem Phys. 2003;118:792–806.
15. Delale CF, Okita K, Matsumoto Y. Steady state cavitating nozzle flows with nucleation. ASME J Fluids Eng. 2005;127:770–7.
16. Brennen CE. Fission of collapsing cavitation bubbles. J Fluid Mech. 2002;472:153–66.
17. Delale CF, Tunç M. A bubble fission model for collapsing cavitation bubbles. Phys Fluids. 2004;16:4200–3.
18. Blake JR, Gibson DC. Cavitation bubbles near boundaries. Ann Rev Fluid Mech. 1987;19:99–123.
19. Kubota A, Kato H, Yamaguchi H. A numerical study of unsteady cavitation on a hydraulic section. J Fluid Mech. 1992;240:59–96.
20. Nigmatulin RI, Khabeev NS, Nagiev FB. Dynamics, heat and mass transfer of vapor-gas bubbles in a liquid. Int J Heat Mass Tran. 1981;24:1033–44.
21. Prosperetti A, Crum LA, Commander KW. Nonlinear bubble dynamics. J Acoust Soc Am. 1988;83:502–14.

22. Prosperetti A. The thermal behavior of oscillating gas bubbles. J Fluid Mech. 1991;222: 587–616.
23. Delale CF. Thermal damping in cavitating nozzle flows. ASME J Fluids Eng. 2002;124:969–76.
24. Franc JP, Michel JM. Fundamentals of cavitation. Dordrecht: Kluwer; 2004.

Experimental Study of the Inertial Motion of Supercavitating Models

N.S. Fedorenko, V.F. Kozenko, and R.N. Kozenko

Abstract

The paper gives a brief overview of various types of available facilities for the experimental study of the high-speed inertial motion of supercavitating bodies in water. The paper reports the procedure of the experimental studies of high-speed supercavitating models which have been conducted at the Hydrodynamics Laboratory of the Institute of Hydromechanics of the National Academy of Sciences of Ukraine under the direction of Yu.N. Savchenko since 1990. The design philosophy of the electrochemical-catapult model firing system and the motion parameter recording system is described. The paper gives examples of model firing and reports the values of the initial parameters, video-recording data on the motion of a supercavitating model, and motion parameter values for models moving with a system of shock waves.

1 Introduction

Models can be put in high-speed motion through water in a number of ways [1, 2]. The types of existing facilities differ in the method of production of the energy delivered to the model to speed it up.

Thus, a controlled-pressure ballistic chamber was built at the Naval Ordnance Test Station, Pasadena, the USA, in 1951. It serves to study the water entry, water exit, and underwater motion of engineless projectiles. A pneumatic piston catapult system fires models of diameter 50.8 mm and mass up to 530 g from a tube into the chamber with water entry and exit speeds up to 36 and 24 m/s, respectively. The 0.9 m square chamber of length 2.4 m has glass windows on three sides and can be set at an angle of 5–90° to the horizontal. The gas pressure in the chamber

N.S. Fedorenko (✉)
Institute of Hydromechanics of the National Academy of Sciences of Ukraine, Kyiv, Ukraine
e-mail: inesteruk@yahoo.com

I. Nesteruk (ed.), *Supercavitation*,
DOI 10.1007/978-3-642-23656-3_2,

over the water surface can be varied from the vapor pressure to 1.5 atm. Observations are made using stroboscopic photography. To measure the variation of the projectile angular velocity in water entry, use is made of a special camera with optical scanning to compensate for the motion of the image relative to the film.

The more recent ballistic chamber at the California Institute of Technology has an electromagnetic catapult system with controlled atmosphere and allows one to study water entry and exit at different angles with waves on the free surface. Model projectiles of diameter 25.4 mm can be fired (at the center of the chamber) transverse to the water–gas interface up and down. Stainless steel models of diameter 25.4 mm are fired underwater at a speed of about 27 m/s, the speed-up distance being 50 mm. Increasing the energy to 54,000 W·s increases the speed to 130 m/s. The speed-up time can be varied by varying the circuit parameters, and an oscillatory motion can be imparted to the model.

The largest controlled-pressure ballistic chamber is installed at the Naval Ordnance Laboratory in White Oak, Maryland, the USA, and it serves to test engineless models. One powder gun fires into the water models of diameter 76.2 mm and mass 5.05 kg at 900 m/s. The other gun has a barrel of caliber 102 mm. Models of diameter up to 76 mm are loaded in a strong titanium cartridge. The tray is stopped by an aluminum braking nozzle of diameter 80 mm at the end of the barrel while the model continues to fly. All operations: model firing, speed calculation, and photography with the use of flash tubes – are performed automatically by a preset program. The chamber length and width are 30 and 10.5 m, and the water depth is 19.5 m.

The Hydrodynamics Laboratory at the California Institute of Technology has a centrifugal catapult system mounted inside a sealed reservoir with water and a gaseous atmosphere over it. Models are fired in a vertical plane at any desired angle with any angle of attack in the range $\pm 10°$ at any speed up to 75 m/s. The water surface has area 3.6×9.16 m, and the water depth is 3.05 m. This special-purpose facility makes possible a variety of experiments both with self-propelled projectiles and with projectiles moving on inertia.

Experimental facilities to test high-speed inertial models have been built and are currently being built in a number of European and Asian countries too. Some results of foreign experimental studies of supercavitating bodies moving at high speeds are presented in [3–5]. At the Institute of Hydromechanics of the National Academy of Sciences of Ukraine, a firing bench has been in service since 1990.

2 Firing Bench at the Institute of Hydromechanics of the National Academy of Sciences of Ukraine

At the Institute of Hydromechanics of the National Academy of Sciences of Ukraine (IHM of NASU), inertial models are fired using a $2{,}100 \times 2{,}100$ mm water tunnel entrance channel of length 35 m. It has ten pairs of windows for optical observations, which are mounted perpendicular to the model trajectory. To keep the model from flying out of the channel, the windows are recessed and protected by the strong walls of the channel, thus assuring test safety. At the end of the test distance, the model is stopped using a metal shield or an obstacle filled with a soft material such as sand, wood, etc. so that the model may not be damaged in stopping.

Models are fired using an electrochemical catapult, which uses ecologically clean components: water, compressed air, hydrogen, and electric current and provides high firing energy at transonic speeds (the sound speed in water at $T = 6°C$ is 1,440 m/s). The action of the electrochemical (gas–vapor) catapult is described and estimates of firing efficiency are given in [6]. The firing bench comprises a hydraulic, a pneumatic, an electric, and a measuring system.

2.1 Hydraulic System

The hydraulic system (Fig. 1) serves to fill the water tunnel channel 10 with water from a basin 7 through a pressure pipe 6 using a pump 2 and to empty the channel 10 through a drain pipe 5 after the experiment. Valves 3 and 4 control the pressure and the drain pipe, respectively.

2.2 Pneumatic System

A schematic of the pneumatic system is shown in Fig. 2 where: 1, 2, 3, 4, 5, 7 – valves, 6 – pressure gage, 8 – gas release to the atmosphere, 9 – compressed-hydrogen

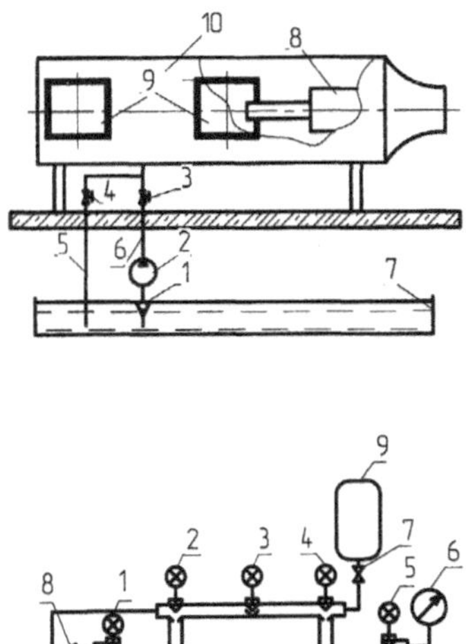

Fig. 1 Schematic of the hydraulic system of the firing bench:.1 – check valve; 2 – pump; 3, 4 – valves; 5 – drain pipe; 6 – pressure pipe; 7 – water basin; 8 – catapult; 9 – observation windows; 10 – water tunnel

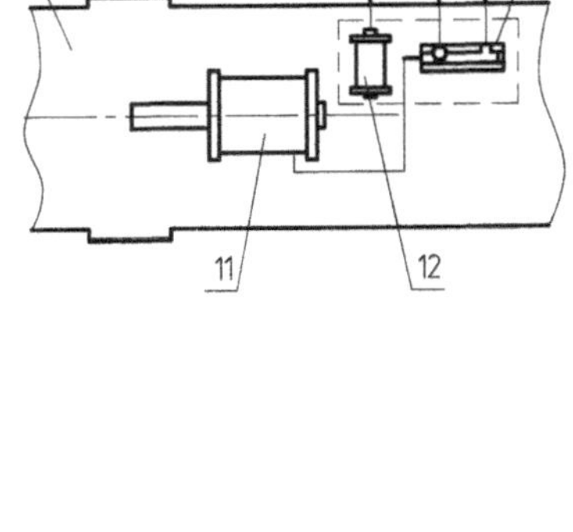

Fig. 2 Schematic of the pneumatic system of the firing bench

bottle, 10 – water tunnel section, 11 – catapult, 12 – electrolyzer, and 13 – check valve block.

2.2.1 Operational Procedure for the Pneumatic System

Before filling the catapult combustion chamber with the products of electrolysis, blow through the system, for which purpose close the valve 1 with the valves 2, 3, 4, 5, and 7 open. Then close the valve 7 and open the valve 1 to complete the blow-through. In doing so, check the pressure on the pressure gage 6.

To fill the catapult combustion chamber with the combustible mixture, close the valve 1 with the valve 2, 3, 4, and 5 open and the valve 7 closed. Turn on the electrolyzer 12 and raise the pressure in the catapult combustion chamber to its working value; in doing so, check the pressure on the pressure gage 6. Once the working pressure is reached, turn off the electrolyzer 12, open the valve 1, and close the valve 5. When the whole of the pneumatic system is vented to the atmosphere 8, the pressure in the catapult combustion chamber remains unchanged due to the check valve 13.

2.3 Electric Circuit

The electric circuit of the bench is shown in Fig. 3 where: 1 – personal computer; 2 – control panel; 3 – video camera; 4 – power unit; 5 – DC generator; 6 – fuse wire; 7 – electrolyzer; 8 – catapult; 9 – window; and 10 – lighting.

The electric system serves to accumulate the firing energy by water electrolysis, fire the gas mixture, start the catapult, and record the test data.

2.3.1 Operational Procedure for the Electric System

To fire a model, fill the catapult combustion chamber with the electrolysis gas. To do so, apply to the electrolyzer a stable working direct current of 80–100 A and a stable working voltage of 11 $\div$ 14.2 V using the power unit 4 and the DC generator 5. Once the required pressure in the catapult chamber is reached, stop the electrolysis. The model is fired and the data are recorded using the control panel 2, the video camera 3, the lighting 10, and the personal computer 1.

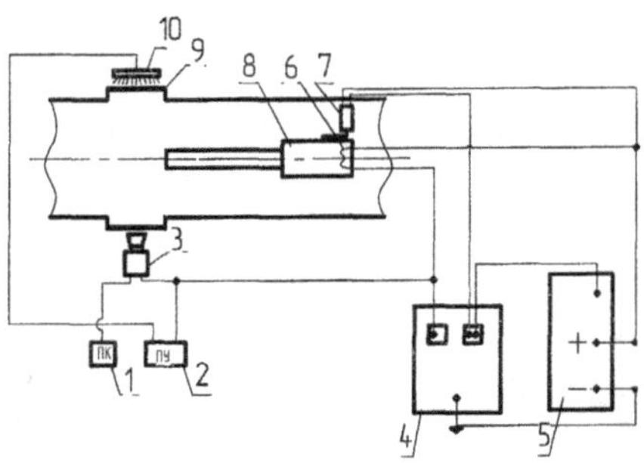

Fig. 3 Electric circuit

The current and the voltage are checked on the amperemeter and voltmeter of the control panel 2. The electric circuit also provides for the synchronous operation of the personal computer 1, the control panel 2, the video camera 3, and the lighting 10 when recording the test data.

The energy is accumulated using the electrochemical process of water decomposition into oxygen and hydrogen by the familiar chemical reaction [7]:

$$2H_2O \rightarrow H_2 \uparrow + O_2 \uparrow .$$

In the process, oxygen is liberated at the anode, and hydrogen is liberated at the cathode. According to Faraday's law, the mass of the oxygen and hydrogen produced at the electrodes will be

$$M = Z \cdot I \cdot t,$$

where $Z_H = 0.0376$ g·/(A · h) and $Z_o = 0.2984$ g·/(A · h) are the electrochemical equivalents of hydrogen and oxygen; I is the current (A); and t is the electrolysis time (h).

Since the produced gases are compressed to pressure P_0 in the combustion chamber, the amount of the accumulated energy can be estimated as

$$E = M_{\mathrm{H2O}} \cdot \Delta H_{289}/18.02 + P_0 V_0,$$

where $\Delta H_{289} = 241.83$ kJ/mole is the water formation heat [kJ/mole] at 289 K (25°C) [8]; M_{H2O} is the water (water vapor) mass in grams; V_0 [m^3] is the combustion chamber volume; and P_0 [Pa] is the combustion chamber pressure prior to firing.

The consumed energy will be

$$E_\Sigma = I \cdot U \cdot t,$$

where I is the circuit current, U is the circuit voltage, and t is the chamber charging (electrolysis) time.

2.4 Measuring System

The measuring system allows one to check the catapult charging parameters. They are the electrolysis current, voltage, and time and the chamber pressure prior to and after electrolysis. The measuring system also records the model motion in the channel using a system of sensors and high-speed photography. Initially, SKS-1 M and Pusk-16 high-speed 16-mm movie cameras with a frame frequency up to 5,000 frames/s were used for this purpose. Now we use an X-Sheam XS4 video camera (Integrated Design Tools, Inc.) with a frame frequency of 1,000–20,000 frames/s.

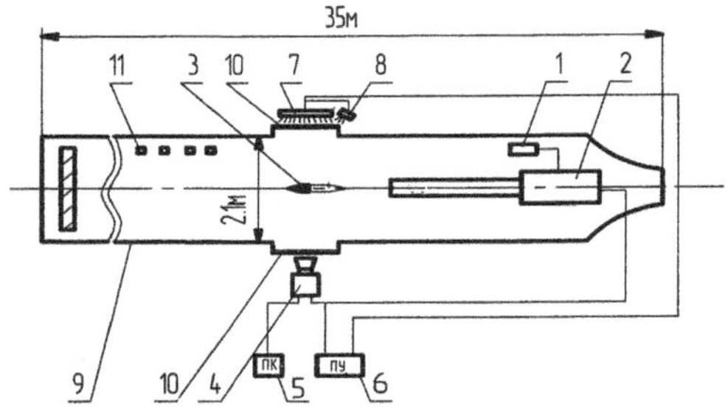

Fig. 4 Schematic of the experimental setup with a video recording

The rather large illuminated area (0.8 × 0.8 m) requires a high lighting power of about 10 kW. Figure 4 shows a schematic of the experimental setup with video recording where: 1 – gas generator; 2 – catapult; 3 – moving model; 4 – video camera; 5 – personal computer; 6 – control panel; 7, 8 – lighting; 9 – water wind channel; 10 – window; and 11 – sensors. The instantaneous speed of models is measured by two methods:

- From the recorded video frames by the technique described in [9]
- By measuring the time it takes for the model to travel the distance between two measuring planes

3 Model Design

Test models must be designed to suit the following basic requirements:

- A model must fit into the supercavity contour over a distance of 35 m
- Stable motion of a supercavitating model over a distance of 35 m
- Minimum deviation of a model from a straight-line trajectory over a specified distance
- Strength sufficient to withstand the accelerating pulse in firing and the longitudinal impact load in water entry
- Strength and stiffness sufficient to withstand the side forces caused by hydrodynamic interaction between the model and the cavity walls (Fig. 5)

To fit a model into the supercavity contour, use is made of the SC_Design program developed at the IHM of NASU [10, 11]. The program constructs the supercavity contour from given parameters: the speed V_x, the hydrostatic pressure P_0, the vapor pressure $P_к$ ($P_к(t) = 2{,}337$ Pa at $T = 20$°C), and the cavitator diameter D_n and fits the model contour into it with some gap between them.

The inertial force F_i and stresses σ_i acting on a model during its acceleration in the barrel to speed V_0 can be estimated as

$$\sigma_i = \frac{4F_i}{\pi D_m^2} = \frac{4ma}{\pi D_m^2} = \frac{2mV_0^2}{\pi D_m^2 L},$$

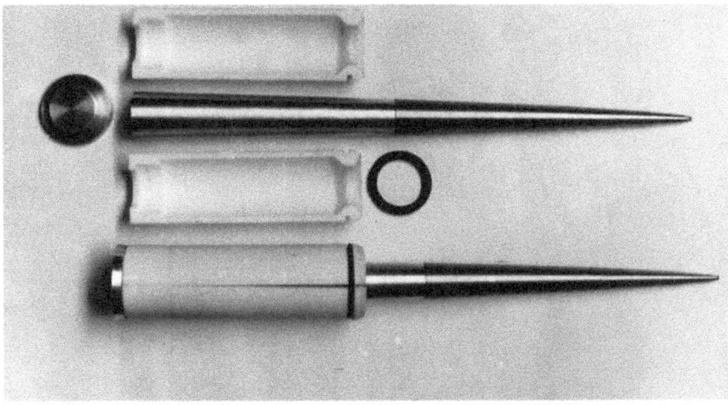

Fig. 5 High-speed supercavitating model

where m is the model mass (kg); $a = \frac{V_0^2}{2L}$ is the acceleration; L_c is the barrel length; and D_m is the model aft diameter.

The test results listed in Table 1 show that in acceleration the model bottom develops stresses of 235–785 MPa. Such high stresses call for special steels with ultimate stresses of the order of 500–800 MPa.

The hydrodynamic drag force F_n and stresses σ_n acting on the cavitator of a model can be estimated as

$$\sigma_n = 4F_n/\pi D_n^2 = C_x \cdot \rho V_0^2/2$$

where $C_x = 0.82$ is the drag coefficient of a disc in a supercavity flow, ρ [kg/m^3] is the water density, V_0 is the model speed, and F_n is the drag force. According to the attained speeds (Table 1), the cavitator stresses will lie in the range 400–910 MPa, which also calls for special high-strength steels.

4 Test Results

Systematic tests on the IHM of NASU's firing bench have been conducted since 1990. Over this period, the following has been investigated:

- The unsteady processes of high-speed water entry and supercavity inception
- The mechanisms of interaction of high-speed supercavitating models with various obstacles
- The features of interaction between high-speed supercavitating models in group motion

Starting in 1993, the obtained results have been published in Refs. [12–20]. Below are some of the test results obtained on the IHM of NASU's upgraded firing bench (see Table. 1). Table 1 gives the catapult charging parameters and the model speeds calculated from the recorded video data for a series of tests. Among the firings shown in the table, of especial interest is firing No 6 because in this case the water sound speed $a = 1{,}422$ m/s at water temperature $T = 4°$C was exceeded.

Figure 6 shows video frames of the motion of the supercavitating model, wherein the supercavity shape and shock waves can be seen.

Table 1 The catapult and the model parameters in experiments

No	Model mass, m_1,kg	Charge mass, m_2, kg	Pressure *P*, MPa	Time *t*, s	Current *I*, A	Voltage, *V*, V	Power, *N*, kW	Speed, V_1, m/s	Acceleration, *a*, m/s²	Barrel length, *L*, m	Kinetic energy, E_k, J
1	0.015	0.032	18	6,60	80	11.0	1.51	1,210	366,025	2.0	2,412.8
2	0.015	0.035	17	6,28	90	11.5	1.78	1,205	363,006	2.0	2,577.8
3	0.015	0.035	17	9,78	90	11.5	2.8	1,205	363,006	2.0	2,577.8
4	0.035	0.035	18.7	6,78	80	10.8	1.65	1,170	344,176	2.0	8,552.8
5	0.014	0.032	17	7,80	100	11.0	2.76	0,955	227,380	2.0	
6	0.015	0.065	11	4,80	100	12.7	1.69	1,550	600,625	2.0	8,028.8
7	0.015	0.068	13	6,90	100	12.8	2.45	1,330	442,225	2.0	6,137
8	0.015	0.068	15	6,00	100	12.7	2.16	1,240	384,400	2.0	5,334.5
9	0.015	0.066	11	5,10	100	12.0	1.7	1,380	476,100	2.0	6,412.8
10	0.015	0.071	11.5	5,10	100	13.1	1.85	1,350	455,65	2.0	6,573.9
11	0.014	0.075	11.5	5,10	100	13.0	1.84	1,205	363,006	2.0	5,593.3
12	0.015	0.072	11.5	5,40	100	13.0	1.95	1,300	422,500	2.0	663.4
13	0.015	0.033	11.5	2,28	100	13.0	0.82	1,200	214,925	3.35	2,424.5
14	0.015	0.060	11.5	4,80	100	14.2	2.08	1,350	272,015	3.35	5,625.5
15	0.016	0.084	11.5	4,95	100	14.2	1.96	1,375	282,183	3.35	8,102.6
16	0.015	0.103	11.5	4,50	100	13.3	1.65	1,230	226,806	3.35	8,053.3
17	0.036	0.080	12.7	5,46	100	13.0	1.97	1,320	260,060	3.35	7,111.8
18	0.015	0.123	11.5	4,80	100	13.0	1.73	1,270	240,731	3.35	10,1217

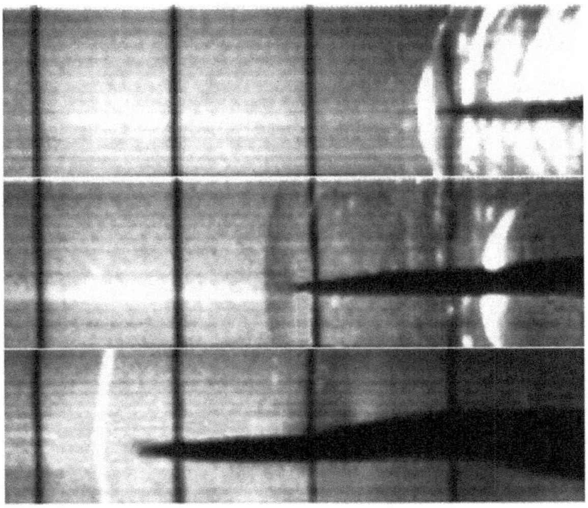

Fig. 6 Record of supersonic motion through water (video frames)

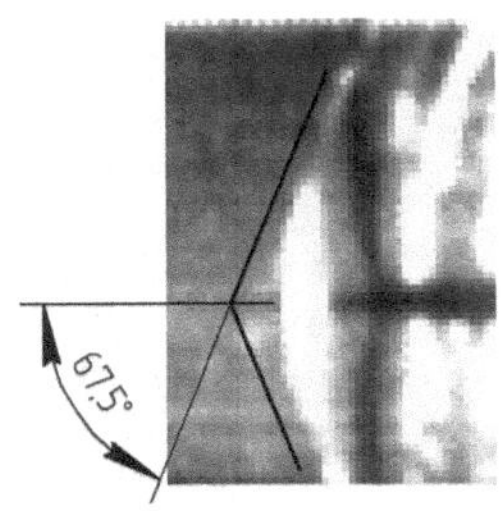

Fig. 7 Mach angle determination

The experiment was conducted under the following conditions:

Ambient parameters:

- Water temperature $T = 4°C$
- Water sound speed $a = 1{,}422$ m/s
- Model immersion depth $H = 0.5$ m

Model parameters:

- Cavitator diameter $D_n = 1.2$ mm
- Model length – 85 mm

Recording parameters:

- Frame frequency 25,000 frames/s
- Exposure time 1 μs
- Graticule scale spacing 50 mm

The model speed can be found from the recorded frames (Fig. 6) [9] and from the shock wave shape (Fig. 7) [20].

The model speed measured from the frames was $V = 1{,}550$ m/s, and the attained Mach number was:

$$M = \frac{V}{a} = \frac{1550}{1422} = 1.09.$$

From the shock wave shape in Fig. 7, the Mach number was estimated as [19]

$$M = \frac{1}{\sin\ 67.5^\circ} = 1.082.$$

References

1. Knapp R, Daily J, Hammitt F. Cavitation (in Russian). Moscow: Mir Publishers; 1974.
2. Gorshkov AS, Rusetsky AA. Cavitation tunnels (in Russian). Leningrad: Sudostroenie; 1972.
3. Kirscner IN. Results of selected experiments involving supercavitating flows. VKI/RTO Special Course on Supercavitation. Brussels: Von Karman Institute for Fluid Dynamics; 2001.
4. Hrubes JD. High-speed imaging of supercavitating underwater projectiles. Exp Fluids. 2001;30(1):57–64.
5. Schaffar M, Ray C, Boeglen G. Behaviour of supercavitating projectiles fired horizontally in a water tank: theory and experiments. 35th AIAA Fluid Dynamic Conference and Exhibit; 6–9 June 2005. Toronto; 2005.
6. Deinekin YuP. Firing of bodies using a gas–vapor catapult (in Russian). Gidromekhanika. 1993;66:40–4.
7. Goronovsky IT, Nazarenko YuP, Nekryach EF. Chemistry handbook (in Russian). Kiev: Naukova Dumka; 1974.
8. Yavorsky BM, Detlaf AA. Physics handbook (in Russian). Moscow: Nauka; 1974.
9. Konovalov NA, Lakhno NI, Putryk ND, Skorik AD. Still and motion picture photography methods in technical mechanics (in Russian). Kiev: Naukova Dumka; 1990.
10. Savchenko YuN, Semenenko VN, Putilin SI, Naumova EI. Software system to simulate the motion of supercavitating bodies in water (in Russian). Matematicheskie Mashiny i Sistemy. 1999;2:48–57.
11. Semenenko VN. Software for designing the supercavitating vehicles. Proceedings of the 10th International Scientific School "High Speed Hydrodynamics (HSH-2008)"; 10–14 September 2008, Cheboksary; 2008. p. 241–52.
12. Savchenko YuN, Semenenko VN, Serebryakov VV. Experimental study of developed cavity flows at subsonic flow velocities (in Russian). Doklady AN Ukrainy. 1993;2:64–9.
13. Savchenko YuN, Semenenko VN, Serebryakov VV. Experimental verification of asymptotic formulas for axisymmetric cavities at $\sigma \to 0$ (in Russian). Problems in high speed hydrodynamics. Cheboksary: Chuvash University; 1993. p. 225–30.
14. Savchenko YuN, Vlasenko YuD, Semenenko VN. Experimental investigations of high-speed cavity flows (in Russian). Gidromekhanika. 1998;72:103–11.
15. Vlasenko YuD. Experimental investigations of high-speed unsteady supercavitating flows. Proceedingsof the Third International Symposium on Cavitation. Vol. 2. Grenoble; 1998. p. 39–44.
16. Savchenko YuN, Morozov AA, Savchenko VT, Semenenko VN. Mathematical models of the motion of supercavitating bodies in water at transonic speeds and systems for its implementation (in Russian). Matematicheskie Mashiny i Sistemy. 1999;1:3–15.

17. Vlasenko YuD. Experimental investigations of supercavitation flows at subsonic and transonic velocities. Proceedings of the International Summer Scientific School "High Speed Hydrodynamics"; June 2002. Cheboksary; 2002. p. 197–204.
18. Savchenko YuN, Semenenko VN, Putilin SI, et al. Designing the high-speed supercavitating vehicles. Proceedings of the 8th International Conference on Fast Sea Transportation (FAST'2005); 27–30 June 2005, St. Petersburg; 2005. ISBN 5-88303-045-9.
19. Savchenko YuN, Semenenko VN, Putilin SI, et al. Some problems of the supercavitating motion management. Sixth International Symposium on Cavitation CAV2006; September 2006. Wageningen; 2006.
20. Savchenko YuN, Zverkhovsky AN. Procedure of experimental study of the high-speed motion of supercavitating inertial models in water (in Russian). Prikladnaya Gidromekhanika. 2009;11(4):69–75.

Two Dimensional Theory of Cavitation Flows Under Ship Bottoms

Michael V. Makasyeyev

Abstract

The two dimensional theory of cavity flows under ship bottoms is based on the linear theory of wave motions of ideal incompressible fluid. The cases of high speed and displacement ships are considered. In the case of high speed ship the problem of planing hull with step and cavity at free fixing of trim angle and draft, unknown shape and length of cavity and wetted borders of hull is solved. The possibilities of modeling of ship hydrodynamic characteristics changing with the help of cavity pressure control are shown. A reduction of the wave resistance can be a result of such changes. In the case of displacement ship the cavitation flow model behind wedge under solid wall is considered. It is shown that the gravity waves with decreasing amplitude on cavity boundary are generated if the cavity on the horizontal wall is closed. In theoretical model, the existence of countable number of cavity lengths is possible. The characteristics of cavity shapes at negative cavitation numbers are determined.

1 Introduction

The cavity flows under ship bottom are created in the purpose to reduce the drag and to control the hydrodynamic characteristics (see examples [1–3]). The special steps for a cavity creation on high speed hulls are designed. The cavities occur in the areas behind steps as a result of high speed motion or pumping of air. On the bottoms of displacement type of ships the artificial ventilated cavities are created essentially. It is possible to control pressure distribution on bottom and make the additional pressure or underpressure. In addition, the hull constructions with steps, air cavities and controllable angles of installation allow obtaining the new

M.V. Makasyeyev (✉)
Institute of Hydromechanics of Ukrainian National Academy of Sciences, Kyiv, Ukraine
e-mail: M_Makasyeyev@ukr.net

I. Nesteruk (ed.), *Supercavitation*,
DOI 10.1007/978-3-642-23656-3_3, © Springer-Verlag Berlin Heidelberg 2012

hydrodynamic effects associated with wave generation. Such effects can cause the reduction of wave resistance and implementation of academician G. E. Pavlenko's idea [4] about wave energy regeneration in system of planing surfaces when some surfaces use the energy which was generated by other surfaces at their right relative position.

The research of real artificial cavity flow under ship bottom represents sufficiently complicated theoretical and experimental problem. Different aspects of vapor and artificial cavity flows near solid walls were researched in works by R. Knapp [5], G.V. Logvinovich [6], Y.N. Savchenko [7, 8]. The works of A.A. Butuzov [9–12] are dedicated to modeling of cavities under ship bottom. The numerical model in these works is based on the two dimensional linear theory of cavity flow behind wedge under solid horizontal flat wall. The calculation results are compared with experiment which was obtained in a hydrodynamic channel for a cavity under a flat plate with side discs. The description of Butuzov's method can be found in books [13, 14]. Later this method was also used by K. Matveev, see, for example, [15].

Butuzov has spread his idea of solution method for the problem with a cavity under wall to the modeling cavity on planing surface behind step [11, 12]. His approach is based on simplified model of planing with the use of Ryabushinsky's cavitation scheme for coupling solid and free boundaries. The free boundaries of fluid in this model are represented as linear solid wall and lengths of wetted segments are given. Cavity length, unknown in physical problems, is given in Butuzov's method and cavitation number is defined from the problem solution. It explains that the cavity length in mathematical problem defines the free boundary and makes the problem nonlinear and in this time the cavitaton number enters linearly into the problem.

In reality the cavitation phenomenon on moving ship bottom differ from simplified models. The main fact is that the moving ship position on the water surface cannot be arbitrarily defined. This position is defined by motion speed, displacement amount, mass distribution and shape of hull. If the cavity exists under the bottom, its shape and length are unknown. The sizes of cavity will depend substantially on the pressure in the cavity, i.e., the cavitation number, and other parameters – Froude number, bottom geometry. From physical conditions of cavitation, it follows that the cavitation number is given. It is defined by the saturated vapor pressure for natural vapor cavitation and by the cavity pressure that is artificially created in the case of artificial gas ventilation. Thus, the cavitation number can be positive or negative. Accordingly, the physical effects caused by cavitation will be different. Particularly, at a negative cavitation number additional backup is created which decreases the draft. Consequently, it is possible to control the hydrodynamic characteristics of the high speed and the displacement type ships with the help of steps with cavities behind it and artificial ventilation of hull bottom segments.

The computational methods are need for comprehensive research of cavities influence on hydrodynamic characteristics of hulls and these methods must correspond with physical phenomena. The calculation method for planing hydrofoil that gives possibility to define the wetted length, pressure distribution and trim angle on given displacement, geometry and center of mass position is shown in [16]. This method is suitable for calculation of system of planing surfaces that have segments with cavities which are the boundaries with a known pressure. The general approach to the study of system of planing surfaces is presented in work [17].

The present work consists of two logical parts united by common mathematical theory. The theory is based on reduction of original problem for velocity potential with linear boundary conditions to the singular integral equations. In the first part the two dimensional theory of the planing hull with a cavity behind a step under the bottom is presented. The examples of numerical results that show the model possibilities are given. These results and physical effects are discussed. In second part, the Butuzov's approach to the modeling of the flow under the ship bottom with a cavity behind a wedge under infinite horizontal wall is developed. This approach can be used in the case of the displacement air cavity ship when the cavity area is small in comparison with the hull bottom size. Improvement of this approach that can define the cavity length at giving the cavitation number and the Froude number is presented.

The part of this work – the results related to the problem of step planing hull with cavity was presented on International Symposium on Cavitation CAV2009 in Michigan university (Ann Arbor, USA) and published in proceedings of Symposium [18].

2 Statements of Problems

2.1 Physical Problem of Planing Hull Motion with a Step and a Cavity on the Bottom

The problem of stepped planing boat moving at constant velocity V_0 over an undisturbed surface of an infinitely deep ideal incompressible liquid is considered (Fig. 1). The boat has two surfaces with unknown in advance wetted lengths l_1 and l_2. These amounts have to be found as part of the solution of the problem interacting with the liquid. The distance between the trailing edges of the surfaces is L. The level of the undisturbed liquid surface coincides with the x-axis.

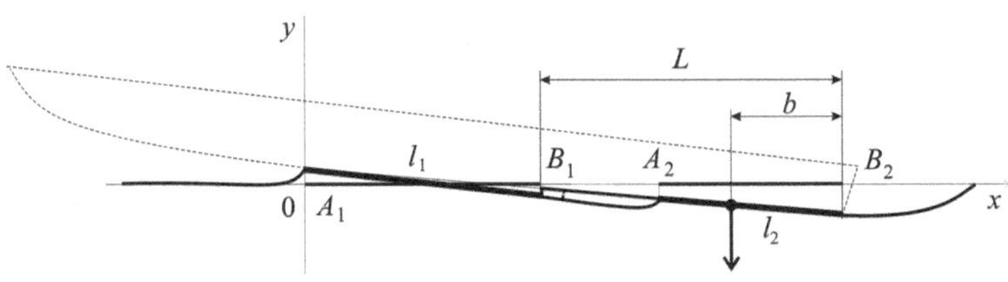

Fig. 1 Scheme of a planing boat with a gas cavity under the bottom after the step

The weight (volume displacement) of the boat is Δ and its center of mass is situated by distance b from the trailing edge of the second planing surface. The values Δ and b are given.

The motion of the boat is modeled by the motion of a system of two flat plates rigidly joined into an integral structure, the x-projections of wetted sections of plates are segments $[A_1, B_1]$ and $[A_2, B_2]$. It is assumed that the angles of the plates with the move direction α_1 and α_2 are small and assumptions of the linearized theory of liquid wave motion are true.

The mathematical model of the physical problem is a boundary-value problem for the perturbed velocity potential, and the boundary conditions are transferred to the axis $y = 0$. On the segments $[A_i, B_i]$, $i = 1, 2$, the unknown pressure difference – the functions $\gamma_i(x) = (p(x, -0) - p_0)/\rho V_0^2$, $x \in (A_i, B_i)$, $i = 1, 2$, are defined where p_0 is the pressure on the free boundary, $p(x, y)$ is the pressure in the liquid, and ρ is the liquid density. The pressure p_c in the cavity aft of the step is specified by the cavitation number $\sigma = 2(p_0 - p_c)/\rho V_0^2$. There is no pressure difference on the free surface at $x < A_1$ and $x > B_2$, and the free surface shape is unknown. The Froude number $Fr = V_0/\sqrt{ga}$ is defined by the characteristic length $a = \sqrt[3]{\Delta/\rho g}$ where g is the gravity acceleration.

It is assumed that the free surface boundary and the plate boundaries represent one streamline. The functions which describe the forms of flow on segments are designated as $f_i(x) = h_i + k_i x$, $i = 1, 2$, h_i are drafts, $k_i = \tan \alpha_i$. The amounts $\Delta\alpha = \alpha_2 - \alpha_1$ and $\Delta h = h_2 - h_1$ (height of step) are given as constructive parameters in conditions of linear approximation.

2.2 Mathematical Problem of Planing Hull Motion with Step and Cavity on Bottom

The boundary-value problem for the velocity potential $\varphi(x, y)$ is as follows:

$$\varphi_{xx} + \varphi_{yy} = 0, \quad y < 0 \tag{1}$$

$$\varphi_y(x, -0) = -\eta_x(x), -\infty < x < \infty, \ x \neq A_1, \ A_2, \tag{2}$$

$$\varphi_x(x, -0) - v\eta(x) = \gamma(x), -\infty < x < \infty, \ x \neq A_1, \ A_2 \tag{3}$$

$$\varphi_x, \ \varphi_y \to 0, \quad y \to -\infty, \tag{4}$$

$$\varphi(-\infty, y) = \varphi_0(x, y), \tag{5}$$

where $\eta(x)$ is the shape of the streamline made up by the free surface boundary and the plate boundaries being flown past, $v = 1/Fr^2$, and

$$\gamma(x) = \begin{cases} \gamma_i(x), & x \in [A_i, B_i], \\ 0, & x < A_1,\ x > B_2, \\ -\dfrac{\sigma}{2}, & x \in [B_1, A_2], \end{cases}$$

is the function that defines the dimensionless pressure difference along the whole liquid surface. Equation 2 is the kinematic condition for smooth flow past the boundary, (3) is the dynamic condition for the pressure on the boundary, which is the Bernoulli equation, (4) is the condition for disturbance attenuation at a great depth, and the condition (5) means that the flow potential is specified at infinity in front of the boat – undisturbed flow or steady-state independent waves.

2.3 Physical Problem of the Cavitation Flow Under the Bottom of a Displacement Ship

The problem of the cavitation flow under the bottom of a displacement ship can be considered as a particular case of the previous problem at zero angles α_1 and α_2. The segment $[A_1, B_1]$ will be presented as polygonal line in view of horizontal semi infinite segment and short segment, which is cheek of edge, under angle of slope.

In this case the axis of abscissas must pass on the level that corresponds with immersion depth of bottom. However, this level can be considered as zero and the pressure and cavitation number needs to be corrected by corresponding addition. The graphic illustration of the physical problem is shown in Fig. 2.

The characteristic length a can also be defined with the use of the ship displacement and some characteristic value, for example, the wedge length c. It is assumed that $|d/c| < < 1$, where d is the wedge height.

At high Froude numbers or at $\nu \to 0$ this problem is equal to the problem of symmetrical cavitation flow of a weightless fluid over an edge. The Froude numbers can be sufficiently small in the case of the displacement ship.

The effects of wave generation on cavity boundary would be expressed strongly with reduction of Froude number. The sufficient long cavities can be generated at intensive air ventilation or gas pumping. It is assumed that there can be long cavities of wave shape whose boundary can intersect the wall level line. It is possible

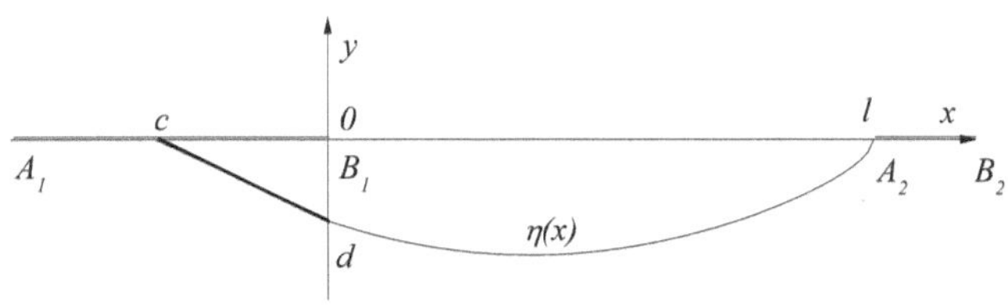

Fig. 2 Scheme of physical statement of the problem of the cavitation flow behind a wedge under the bottom of a displacement ship

physically if a hollow exists on the wall (ship bottom) in area in front of cavity closure point on segment $[B_1, A_2]$, in which the cavity goes into freely without contact of walls.

2.4 Mathematical Problem of Cavity Flow Under Bottom of Displacement Ship

The mathematical problem of cavity flow under bottom can be written formally in the form (1)–(5). In this case the exceptions for conditions (2) and (3) will be the point of flow return on beginning of wedge cheek and the junction point of the cavity and the solid wall. The potential $\varphi_0(x, y)$ is identically equal to zero. It will correspond to the absence of independent steady waves ahead.

Note that boundary conditions on segment $[B_1, A_2]$ correspond to the conditions on the boundary of the cavity where the pressure is given and it is constant and the boundary of the cavity is a part of the stream line. The presence of the wall on this segment or its shape does not matter formally.

3 Solution Method for Boundary Problem

3.1 Common Theory

The problem (1)–(5) is solved using the Fourier method for the construction of fundamental solutions [19]. After a conversion to the generalized functions in (1)–(5) and a construction of the fundamental solution of the Laplace generalized equation it is possible to obtain the relationship [20] with the use of the boundary conditions (2)–(3):

$$(|\lambda| - \nu)H(\lambda) = \Gamma(\lambda), \qquad (6)$$

where $H(\lambda) = F[\eta(x)](\lambda)$ and $\Gamma(\lambda) = F[\gamma(x)](\lambda)$ are generalized Fourier transforms of functions $\eta(x)$ and $\gamma(x)$ respectively. The functional relations (6) connect the generalized Fourier transforms of pressure function and boundary form function. The velocity potential as an auxiliary function is excluded.

The inverse Fourier transformation gives

$$\frac{1}{\pi}\int_{-\infty}^{\infty}\frac{\eta(s)}{(x-s)^2}ds + \nu\eta(x) = -\gamma(x), -\infty < x < \infty, \qquad (7)$$

where the singular integral exists in the meaning of Hadamard. The relationship (7) is true in generalized functions. The classical functions $\eta(x)$, $\gamma(x)$ and these derivatives are sectionally continuous functions and if the boundary form $\eta(x)$ or this derivative are known on some section then pressure function $\gamma(x)$ is unknown on this section. The opposite is also true. If the pressure is known in some section then the form function is unknown there. Equation 7 gives the integral equation for problems where the pressure is known on the part of the boundary and the shape function is finite. The integral will be replaced by an integral with finite limits in this case.

For other problems it is necessary to have expression for function $\eta(x)$ per $\gamma(x)$, i.e., it is necessary to resolve the Eq. 7. The resolution of (7) can be found with the use of the fundamental solution construction by Fourier method. Thus, the formal solution of Eq. 6 in generalized functions should be found. The generalized function

$$H(\lambda) = \Gamma(\lambda)\left(reg\frac{1}{|\lambda| - v}\right) + [A\Gamma(v) + A_0]\delta(\lambda - v) + [B\Gamma(-v) + B_0]\delta(\lambda + v), \tag{8}$$

satisfies this equation. Here *reg* indicates regularization, A, B, A_0, B_0 are arbitrary complex constants, $\delta(\lambda)$ is delta function. The constants A and B are defined from conditions on infinity and A_0, B_0 define the homogeneous solution at $\Gamma(\lambda) = 0$ that corresponds to $\varphi_0(x, y)$ in (5).

The inverse transformation of (8) is

$$\eta(x) = \int_{-\infty}^{\infty} \gamma(s)Q(v, x - s)ds + \frac{A}{2\pi}\int_{-\infty}^{\infty} \gamma(s)e^{-iv(x-s)}ds + \frac{B}{2\pi}\int_{-\infty}^{\infty} \gamma(s)e^{iv(x-s)}ds + \frac{A_0}{2\pi}e^{-ivx} + \frac{B_0}{2\pi}e^{ivx} \tag{9}$$

where

$$Q(v, x) = F^{-1}\left[reg\frac{1}{|\lambda| - v}\right] = -\frac{1}{\pi}\left[\cos vx\mathrm{Ci}v|x| + \sin v|x|\left(\frac{\pi}{2} + \mathrm{Si}v|x|\right)\right],$$

Si and Ci are the integral sine and cosine.

Let us write the condition of waves absence on ahead infinity in front on hull and wedge at zero homogeneous solution at $A_0 = 0$ and $B_0 = 0$. Since $\lim_{|x|\to\infty} \mathrm{Si}v|x| = \pi/2$ and $\lim_{|x|\to\infty} \mathrm{Ci}v|x| = 0$, then

$$\lim_{|x|\to\infty} Q(x, v) = -\sin v|x|. \tag{10}$$

Hence,

$$\eta(x)|_{x\to\pm\infty} = \int_{-\infty}^{\infty} \gamma(s)\sin v|x-s|ds + \frac{A}{2\pi}\int_{-\infty}^{\infty}\gamma(s)e^{-iv(x-s)}ds + \frac{B}{2\pi}\int_{-\infty}^{\infty}\gamma(s)e^{iv(x-s)}ds$$
$$= \left(\frac{A}{2\pi}\pm\frac{i}{2}\right)\int_{-\infty}^{\infty}\gamma(s)e^{-iv(x-s)}ds + \left(\frac{B}{2\pi}\mp\frac{i}{2}\right)\int_{-\infty}^{\infty}\gamma(s)e^{iv(x-s)}ds. \tag{11}$$

It would be no waves on the left in infinity if one supposes that $A = \pi i, B = -\pi i$. Therefore, (9) can be written as follows:

$$\eta(x) = \int_{-\infty}^{\infty} \gamma(s)[Q(v, x-s) - \sin v(x-s)]ds + a_0 \sin vx + b_0 \cos vx. \tag{12}$$

Here a_0 and b_0 are real constants which define the amplitude of independent waves.

The differentiation of (12) gives the equation that corresponds to the boundary condition (2):

$$\frac{1}{\pi}\int_{-\infty}^{\infty}\gamma(s)\left[\frac{1}{x-s} + vR(v, x-s) + v\pi\cos v(x-s)\right]ds$$
$$= -\eta'(x) - v(a_0 \cos vx - b_0 \sin vx), -\infty < x < \infty, \tag{13}$$

where $R(v,x) = \frac{1}{\pi}\left[\left(\frac{\pi}{2}\operatorname{sgn}(x) + \operatorname{Si}v(x)\right)\cos vx - \operatorname{Ci}v|x|\sin vx\right]$.

The Eqs. 7, 12 and 13 are the base for resolution of formulated problems.

3.2 System of Integral Equation for Problem of Planing Hull with Step and Cavity

In the problem of the planing hull, the Eq. 13 is used for determination of the pressure function and (12) is used for determination of the free surface shape. The integrals in (12) and (13) are replaced with the integrals between the finite limits A_1 and B_2 because the pressure is zero outside of $[A_1, B_2]$.

The forces and the moments balance conditions [16] must be added to the Eq. 13:

$$\int_{A_1}^{B_2} \gamma(x)dx = v, \tag{14}$$

$$\int_{A_1}^{B_2} \gamma(x)xdx = v(B_2 - b). \tag{15}$$

The three Eqs. 13–15 resolve completely the main problem of determination of pressure distribution, wetted length and trim angle in case of one single planing hull.

If we have a cavity under bottom there are two rigidly bound planing surfaces with unknown lengths. Therefore, the Eq. 13 will be written as a system of two equations on each of surfaces. As result the follow integral equations system will be obtained on the base of (13)–(15):

$$\begin{aligned}&\frac{1}{\pi}\int_{A_1}^{A_2} \gamma_1(s)K(v,x-s)ds - \frac{\sigma}{2\pi}\int_{A_2}^{B_1} K(v,x-s)ds + \frac{1}{\pi}\int_{B_1}^{B_2} \gamma_2(s)K(v,x-s)ds\\&= -f_1{}'(x) - v(a_0\cos vx - b_0 \sin vx),\ A_1 < x < A_2,\end{aligned} \tag{16}$$

$$\begin{aligned}&\frac{1}{\pi}\int_{A_1}^{A_2} \gamma_1(s)K(v,x-s)ds - \frac{\sigma}{2\pi}\int_{B_1}^{B_2} K(v,x-s)ds + \frac{1}{\pi}\int_{B_1}^{B_2} \gamma_2(s)K(v,x-s)ds\\&= -f_2{}'(x) - v(a_0\cos vx - b_0 \sin vx),\ B_1 < x < B_2,\end{aligned} \tag{17}$$

where $K(v,x) = \frac{1}{\pi}\left[\frac{1}{x} + vR(v,x) + v\pi\cos vx\right]$.

The Eqs. 16 and 17 contain two unknown functions $\gamma_1(x)$ and $\gamma_2(x)$, two unknown constants $l_1 = B_1 - A_1$ and $l_2 = B_2 - A_2$ (wetted lengths), and the unknown trim angle α_1 or α_2 (the second is determined from the rigid geometry of the structure). The condition (14) has the form

$$\int_{A_1}^{A_2} \gamma_1(s)ds - \frac{\sigma}{2}(L - l_2) + \int_{B_1}^{B_2} \gamma_2(s)ds = v. \tag{18}$$

The condition (15) is

$$\int_{A_1}^{A_2} \gamma_1(s)sds - \frac{\sigma}{4}\left[(L + l_1 - l_2)^2 - l_1^2\right] + \int_{B_1}^{B_2} \gamma_2(s)ds = v(L + l_1 - b) \tag{19}$$

The condition of geometrical closure of cavity is

$$\begin{aligned}
&\int_{A_1}^{A_2} \gamma_1(s)\left[\left(Q(v,L+l_1-l_2-s)+\sin v(L+l_1-l_2-s)\right)-\left(Q(v,-s)+\sin v(-s)\right)\right]ds \\
&\quad -\frac{\sigma}{2}\int_{A_2}^{B_1}\left[\left(Q(v,L+l_1-l_2-s)+\sin v(L+l_1-l_2-s)\right)-\left(Q(v,-s)+\sin v(-s)\right)\right]ds \\
&\quad +\int_{B_1}^{B_2} \gamma_2(s)\left[\left(Q(v,L+l_1-l_2-s)+\sin v(L+l_1-l_2-s)\right)\right. \\
&\quad \left.-\left(Q(v,-s)+\sin v(-s)\right)\right]ds= \\
&=-(L-l_2)\cdot tg(\alpha_1+\Delta\alpha)-l_1\cdot tg\alpha_1+\Delta h,
\end{aligned} \tag{20}$$

where $\Delta\alpha=\alpha_2-\alpha_1$ and $\Delta h=h_2-h_1$ are specified as design parameters in the linear approximation.

3.3 System of Integral Equations of Problem of Cavity Under Bottom of Displacement Ship

The relation (7) gives the integral equation for the problem of cavity behind wedge under solid wall:

$$\frac{1}{\pi}\int_{-\infty}^{\infty}\frac{\eta(s)}{(x-s)^2}ds+v\eta(x)=\frac{\sigma}{2},\ B_1<x<A_2. \tag{21}$$

By means of partial integration it is possible to proceed to equation with Cauchy nuclear. As a result we have obtained the equation for function $q(x)=\eta_x(x)$:

$$\frac{1}{\pi}\int_0^l\frac{q(s)}{s-x}ds+v\int_0^x q(s)ds=\frac{\sigma}{2}-v\eta(0)-\frac{1}{\pi}\int_c^0\frac{q_0(s)}{s-x}ds,\ 0<x<l, \tag{22}$$

where l is cavity length, $q_0(x)$ is known derivative of wedge form function. This equation must be satisfied together with closure cavity condition $\eta(l)=0$, or

$$\int_0^l q(s)ds=-\eta(0). \tag{23}$$

The resolution of singular integral equation (22) must be found in class of functions which are finite in the point $x = 0$ and infinity in the point$x = l$. The first condition ensures the continuous and smooth conjugation of wedge and cavity boundaries that corresponds with Cutta-Joukovski condition. Under second condition in the cavity closure area there can be infinity of derived function of cavity shape or elliptical form. The system (22)–(23) describes the symmetrical cavity flow behind wedge at $v = 0$. The analysis of such system can be found in [21].

The linear Ryabushinski's scheme of cavity closure on imaginary wedge was used in works of Butuzov [9, 10]. The Eq. 22 for Ryabushinski's scheme will contain the additional summand in a form of an integral in right side which corresponds to an imaginary wedge and it is necessary to write the Eq. 23 for condition $\eta(l + b_1) = 0$, where b_1 is the length of the imaginary wedge.

4 Numerical Method

The systems of integral equations (16)–(20) and (22)–(23) may be solved by any of the familiar methods of solution of singular integral equations. The key feature is that they are parametrically nonlinear in the unknowns l and l_1, l_2. The use of any numerical method gives a system of algebraic equations of the form

$$AX = B, \tag{24}$$

where the vector X is made up by the unknown values of the functions $\gamma_1(x)$, $\gamma_2(x)$, and α_1, which enter into the system linearly, while the elements of the matrix $A = A(l_1, l_2)$ depend on the unknowns l and l_1, l_2, which create nonlinearity. The vector B is made up by elements from the known values of the right-hand sides of Eqs. 22–23.

For the solution of systems of this type, it turns out to be efficient to use the familiar method [22] of reduction of the problem (24) to the minimum search of quadratic functional. Particularly in the case of planing the search problem has such view:

$$[A(l_1, l_2)X - B]^{\mathrm{T}}[A(l_1, l_2)X - B] \to \min_{l_1, l_2}. \tag{25}$$

In this work, the singular integral equations are solved using the discrete singularity method [23, 24], and the problem (25) is solved using the Nelder–Mead flexible polyhedron method (downhill simplex method) [25].

5 Calculation Results

5.1 Motion of Planing Hull with Bottom and Cavity

It makes sense to relate the scales of values of the variables that are specified in the problem and define the geometry of the planing boat to the scale of the generated waves. Let us estimate the free surface shape from (11) which is generated by delta-function pressure pulse $\gamma(x) = c\delta(x)$ of strength c moving at velocity V_0. In this case $c = \int_{-\infty}^{\infty} \gamma(x)dx = v$ and free surface shape will be (without taking into account independent waves, at $a_0 = 0$, $b_0 = 0$)

$$\eta(x) = v[Q(v,x) + \sin vx].$$

We can obtain

$$\lim_{x\to\pm\infty} \eta(x) = \begin{cases} 0, & x<0, \\ -2v\sin vx, & x>0. \end{cases}$$

This expression allows the scale of the planing-induced waves for the linearized theory to be estimated. For example, at Froude number $Fr = 2$ the wave amplitude will be 0.5 and the wave length will be 25.13, and at $Fr = 1.5$ the amplitude and the wave length will be 0.88 and 14.14, respectively.

With this in mind, the results of calculations at $L = 15$ (space between the trailing edge and the step), $\Delta h = 0.7$(step height), $\Delta\alpha = 0$ (the planing surfaces fore and aft of the step are parallel), and with the center of mass situated distance $b = 10$ from the trailing edge are presented below.

Figure 3a–h show the cavity and free boundary shape for the cavitation number ranging from $\sigma = 0.2$ to $\sigma = -0.0224$ at $Fr = 2$. Bold lines show here and further the free surface, thin line segments show the wetted boundaries of planing hull. The free surface boundary consists of three areas. The left area begins in minus infinity and ends in a zero point, where a y-axis passes – in the contact point of free surface with beginning of the wetted area of planing hull. The middle area is the free boundary of cavity after step and right-hand area is the wake border.

The cavitation number $\sigma = -0.0224$ for parameters mentioned above is close to the value when second wetted length goes to zero. That means that subsequent increase of pressure in cavity can lead to tearing of stream from back edge and subsequent undesirable unsteady effects.

Figure 4a–f show the cavity and free boundary shape for the $Fr = 1.7$ and the same center of mass on the distance $b = 10$ from the trailing edge. In this case the critical cavitation number is close to $\sigma = -0.035$.

Note that not all combinations of design parameters and sizes of planing boat and step allow constructing a physically feasible flow or making the residual of the system (16)–(20) or the value of the goal function in (25) smaller than a preset small positive number. However, for all the results presented in the paper the goal

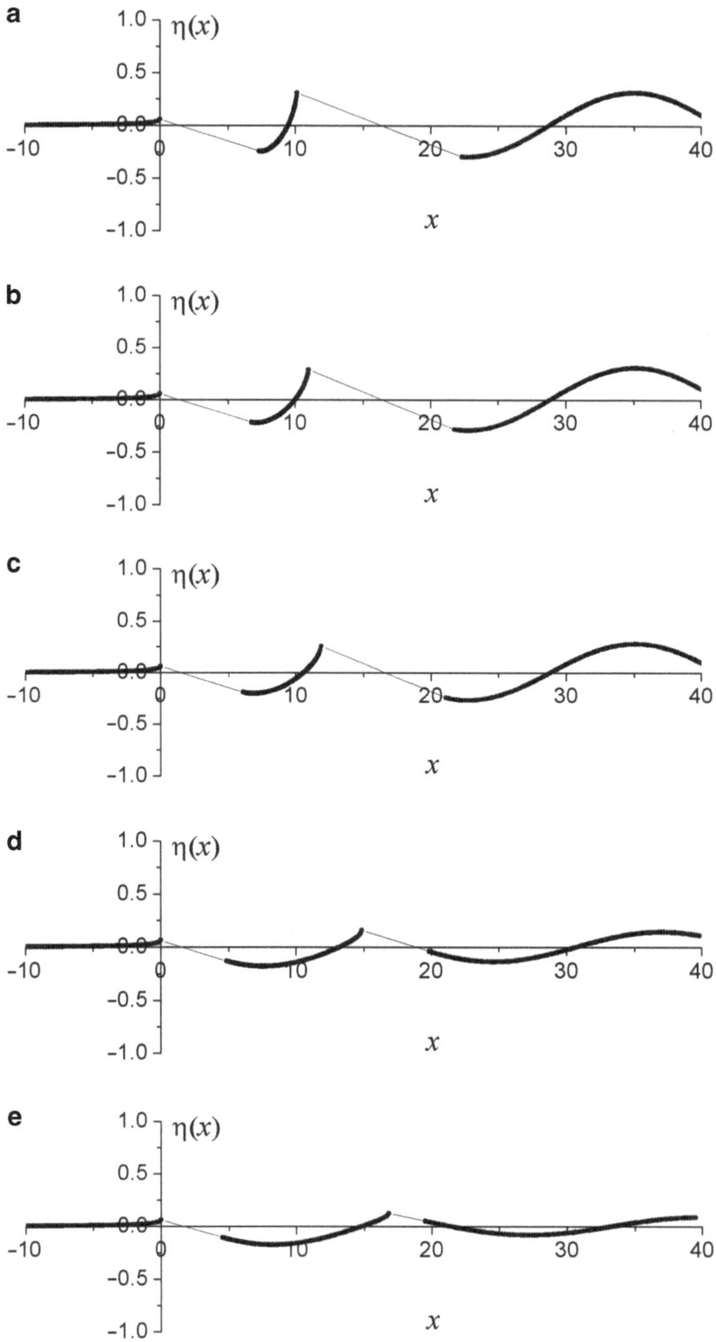

Fig. 3 (continued)

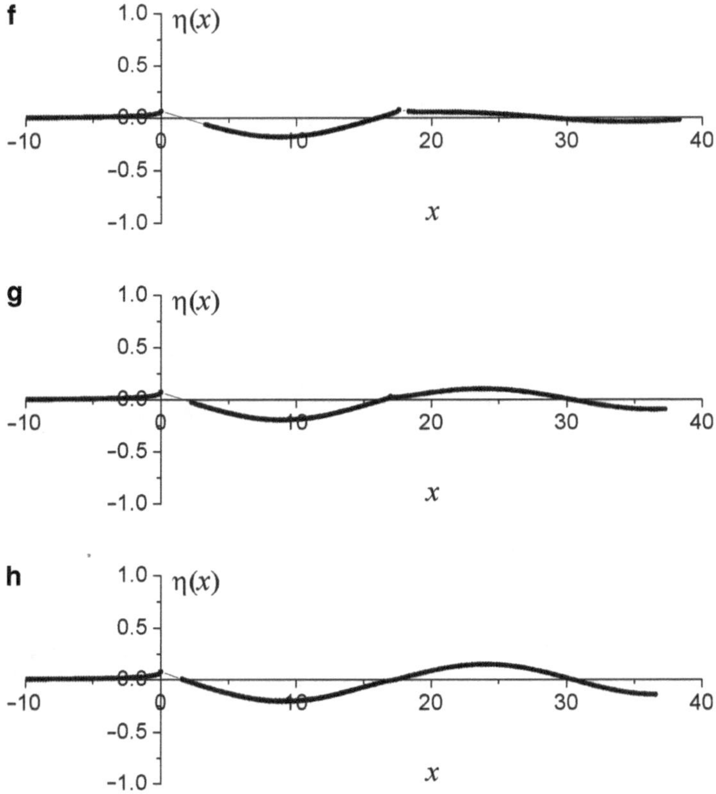

Fig. 3 Shape of the free surface, the cavity aft of the step, and the wetted boundaries of the planing boat at $Fr = 2$. (**a**) $\sigma = 0.2$ (**b**) $\sigma = 0.1$ (**c**) $\sigma = 0.05$ (**d**) $\sigma = 0.0$ (**e**) $\sigma = -0.01$ (**f**) $\sigma = -0.017$ (**g**) $\sigma = -0.02$ (**h**) $\sigma = -0.0224$

function did not exceed 10^{-7}. For the rather small Froude numbers for parameters mentioned above, the flows were constructed for only negative cavitation numbers. The samples of such flow for $Fr = 1.5$ and $Fr = 1.2$ are shown on the Figure 5a–c.

The pressure distributions along the solid boundaries are shown in Fig. 6a–d.

Figure 7 shows the cavity length versus cavitation number at the parameters indicated above. The calculations show that the shape of the cavity behind the step is defined by two factors, namely, by the Froude number and the cavitation number. The Froude number defines the cavity curvature, which correlates with the curvature of the generated waves, and the cavitation number defines the cavity length. At large Froude numbers the cavity curvature is small, and it is increased with the decrease of Froude number. The waves of the same length as the waves in the wake of the planing boat are generated on long cavities at sufficiently small cavitation numbers.

The pattern of contact of the cavity boundary with the second planing surface depends on the cavity curvature in the vicinity of the contact point. The curvature is defined by the ratio between the cavity length and the wave length at a given Froude number. Different contact patterns are illustrated in Figs. 3–5.

At negative cavitation numbers the pressure in the cavity is higher than that on the free surface, and thus an additional lift is developed under the bottom. It can be seen from the plots of the cavity and free boundary shape that at negative cavitation numbers the planing boat draft decreases. The wave amplitude in the wake decreases too. If the second planing surface behind the step is on the trailing wave front, the wake amplitude increases. If the surface is on the leading wave front, the wake amplitude decreases.

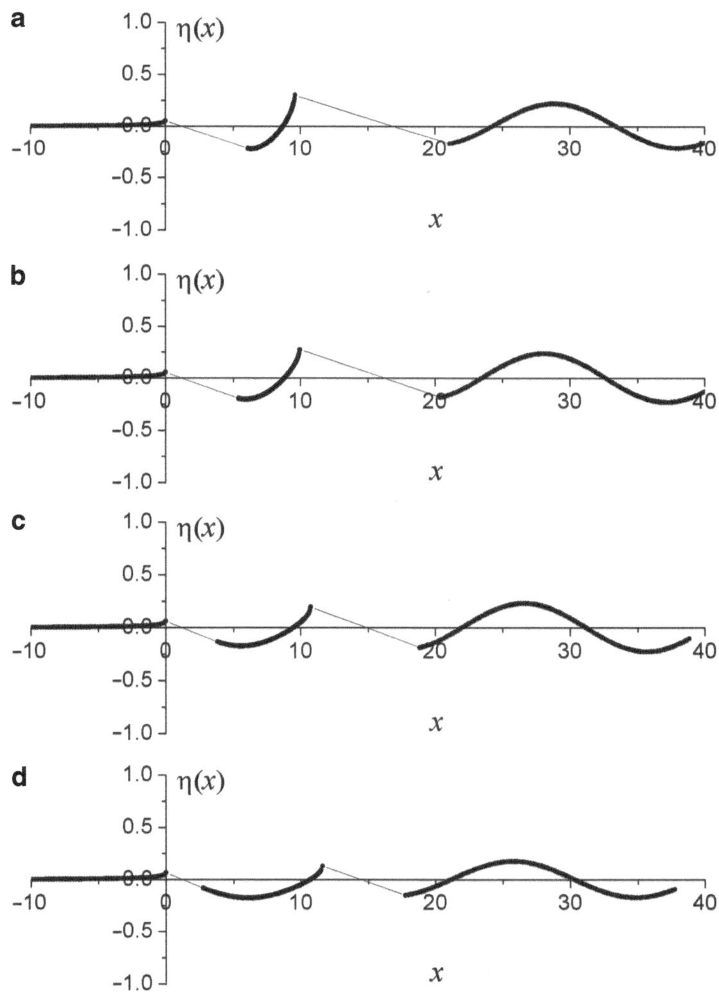

Fig. 4 (continued)

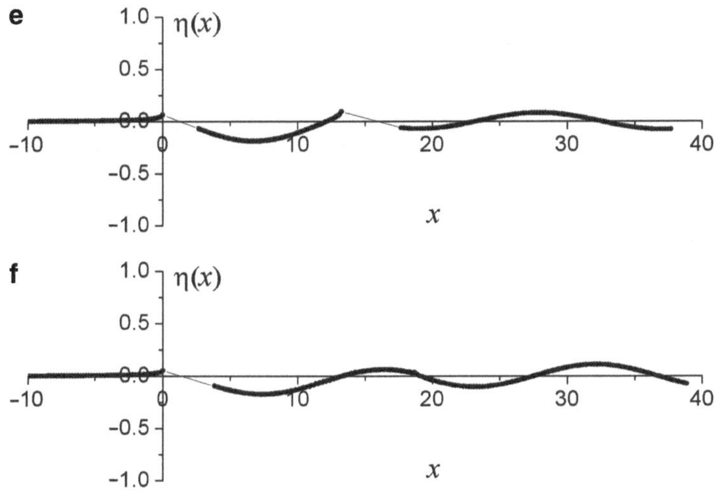

Fig. 4 Shape of the free surface, the cavity behind the step, and the wetted boundaries of the planing boat at $Fr = 1.7$. (**a**) $\sigma = 0.1$ (**b**) $\sigma = 0.05$ (**c**) $\sigma = 0.0$ (**d**) $\sigma = -0.02$ (**e**) $\sigma = -0.03$ (**f**) $\sigma = -0.035$

The analysis of the calculated data shows that one can select an optimum combination of design parameters and factors such that the wave amplitude in the wake is a minimum. The calculated data allow supposing that the consumption of energy to form the wake decreases due to the fact that the second planing surface behind the step uses the energy of the wave generated by the first surface. Multistep planing surfaces with controllable angles of setting can enhance this effect many-fold.

5.2 Cavity Under Bottom of Displacement Ship

It is possible to make some preliminary assumptions about calculation results on accepted model. We will proceed from general theory and known data. In the model of cavity flow behind wedge with finite Froude numbers, the gravity waves are generated on the cavity boundary. The wave length will depend substantially on Froude number. It is logical to suppose that it will be in proportion to value $2\pi/v = 2\pi Fr^2$ which is the wave length on free boundary behind streamlined obstacle. This supposition appeared from the analogous with the case considered in previous section. The cavity length and shape will be determined by cavitation number but will also depend on Froude number.

In model of weightless fluid, the wave length tends to infinity and cavity length depends only on cavitation number. The cavity length is big at zero cavitation number and tends to infinity with Froude number. It is the case of so-called Kirchhoff cavity.

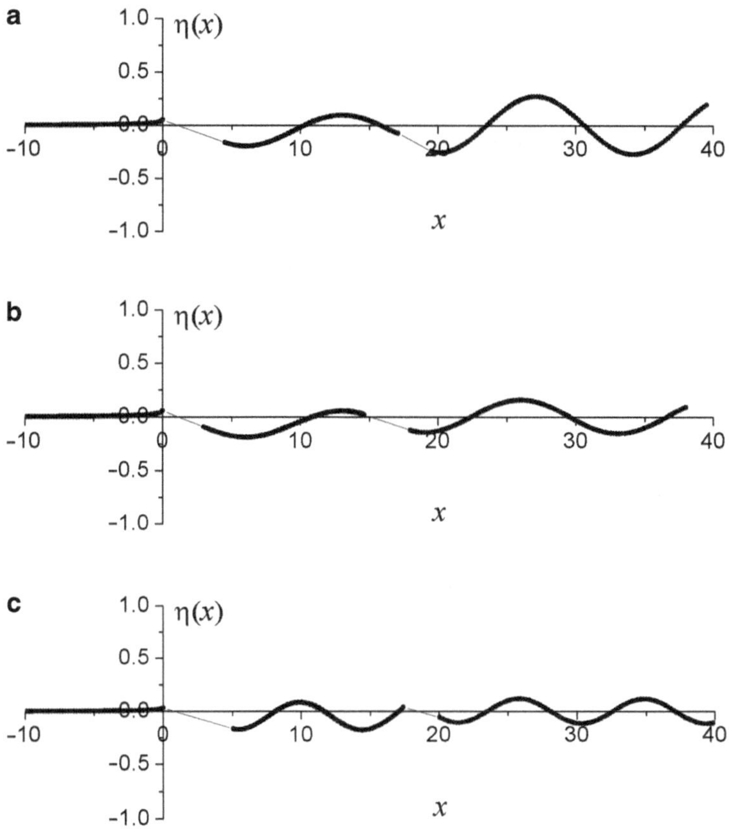

Fig. 5 Shape of the free surface, the cavity behind the step, and the wetted boundaries of the planing boat at $Fr = 1.5$ and $Fr = 1.2$. (**a**) $Fr = 1.5$, $\sigma = -0.04$ (**b**) $Fr = 1.5$, $\sigma = -0.05$ (**c**) $Fr = 1.2$, $\sigma = -0.06$

The cavity length decreases when the cavitation number increases in positive side. The wave length will decrease correspondingly when the cavitation number increases at fixed and finite Froude numbers. As the shape of cavity boundary is wave like at finite Froude numbers, one can suppose that the resolution of problem of cavity length determination at given cavitation number and Froude number will be nonunique. The cavity length will correspond approximately to cross points of wave and solid wall level line behind wedge, i.e. it will be multiple approximately with the half of wave length.

The calculations results confirm these assumptions. The calculations results show that the cavity boundary has wave shape at any Froude number. The cavity length is determined by cavitation number and Froude number from resolution of search problem of minimum (25). The search interval is preset for this problem and corresponds to the expected cavity length. The cavities of minimal length are convex, they do not have bends, and actually, they are half of wave. They are elliptical in the closure area.

The cavity shapes of minimal length behind wedge at $Fr = 7$ and different cavity numbers are presented on Fig. 8. The cavity ordinates and cavity numbers are divided on nondimensional wedge height. It is the case of relatively big Froude number and big length of gravity waves. The dependences of lengths of minimal cavities from Froude number at two values of cavitation number are shown on Fig. 9. The horizontal lines show the limiting values of cavity length at Froude number which tends to infinity. These values correspond to known resolutions for symmetrical cavity wedge in weightless fluid [16].

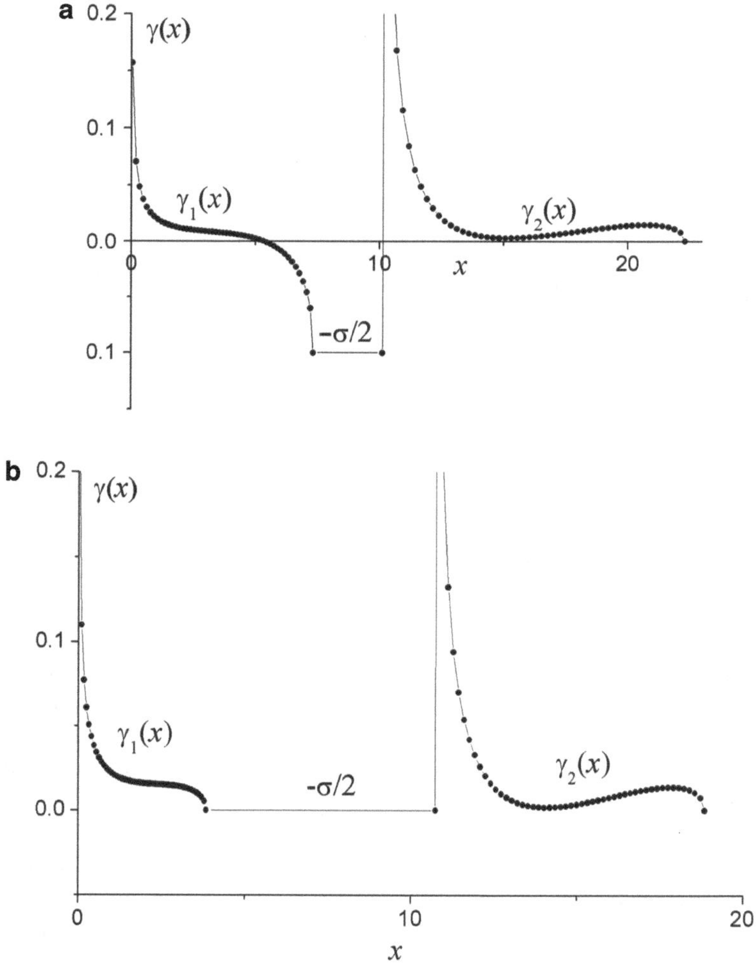

Fig. 6 (continued)

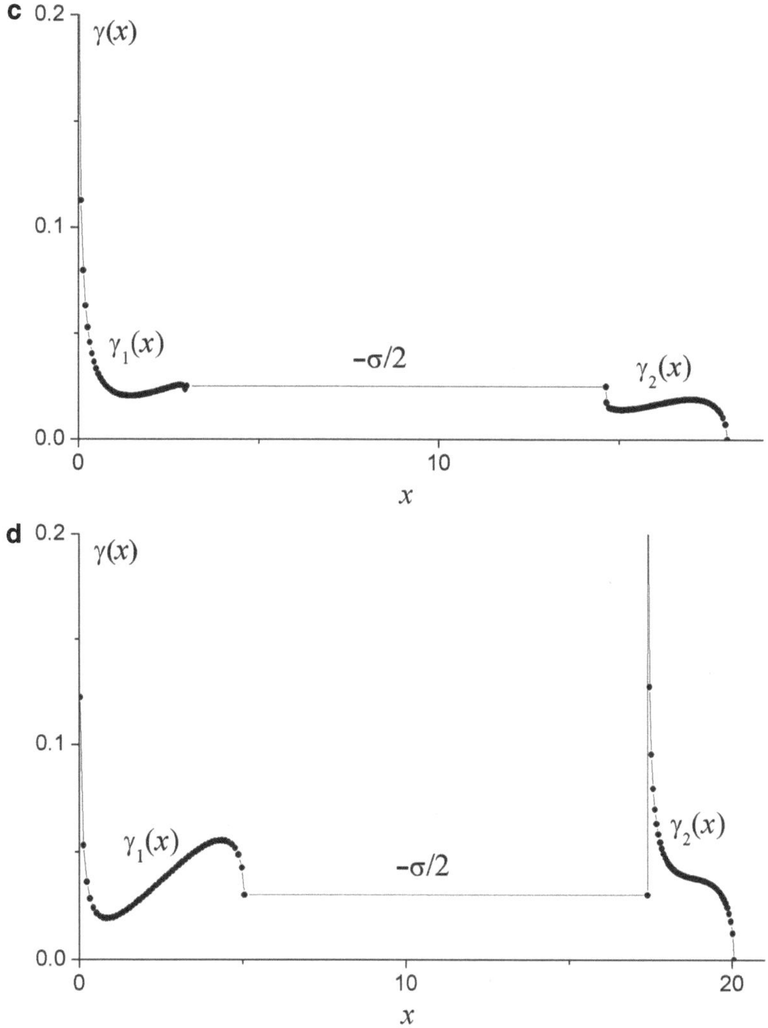

Fig. 6 Pressure distributions along the solid boundaries. (**a**) $Fr = 2$, $\sigma = 0.2$ (**b**) $Fr = 1.7$, $\sigma = 0.0$ (**c**) $Fr = 1.5$, $\sigma = -0.05$ (**d**) $Fr = 1.2$, $\sigma = -0.06$

The calculations of long cavities which length is more than one wave show that the waves amplitudes on cavity boundary are not constants along the length. They are maximal on first wave that comes down from cavitator and they decrease asymptotically to zero on following waves. The cavity shapes at zero cavitaton number $\sigma/|d| = 0$ and Froude numbers $Fr = 3$, $Fr = 1.2$ and $Fr = 0.8$ are presented on Fig. 10a–c respectively.

The obtained results show that the question about cavity length appears in model which takes into accounts the fluid weightiness. For any value couple of Froude

Fig. 7 Calculated cavity length versus cavitation number at $Fr = 2.0$ (*squares*) and $Fr = 1.7$ (*circles*)

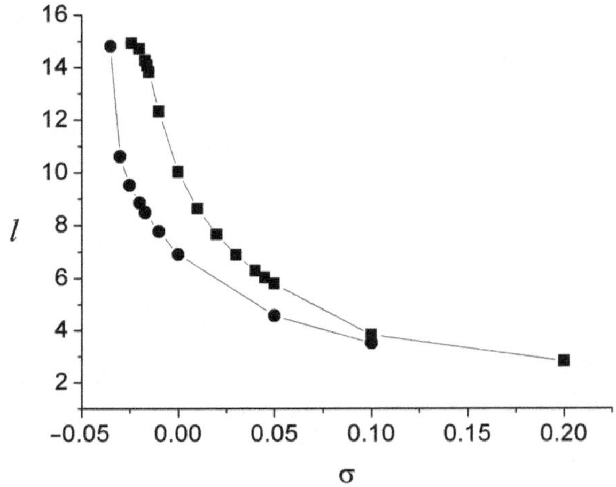

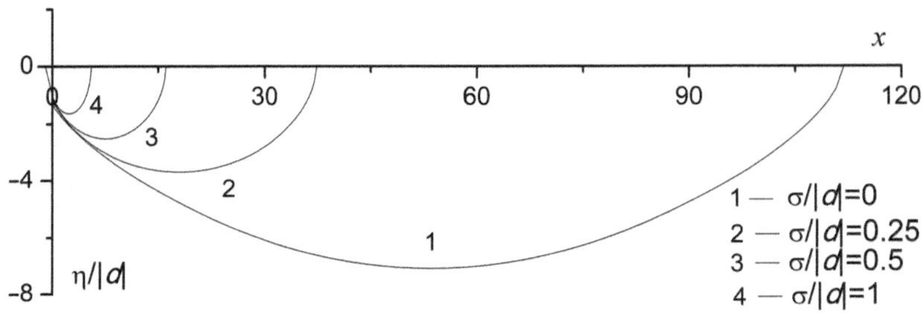

Fig. 8 Cavity shapes behind wedge at $Fr = 7$ and different cavitation numbers

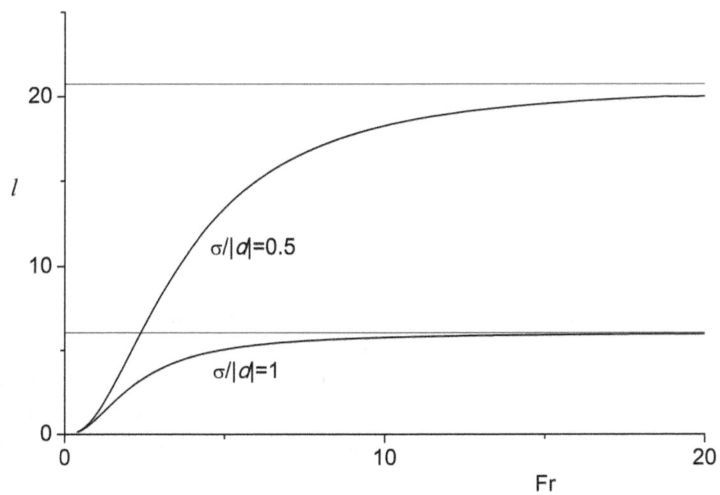

Fig. 9 Dependence of minimal lengths of cavities from Froude number at constants cavitation numbers

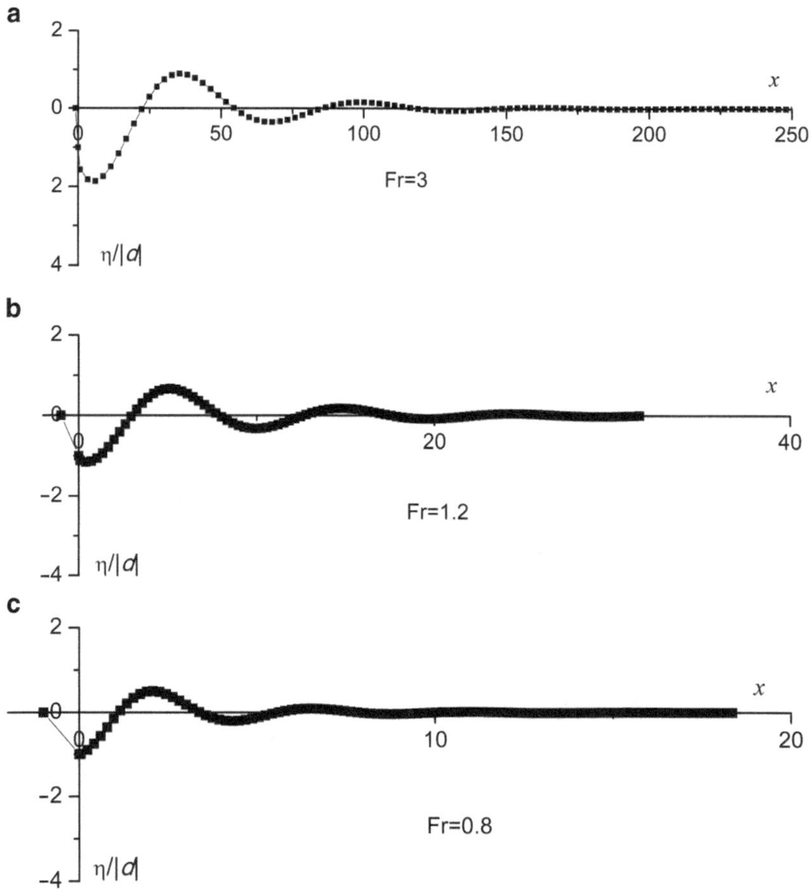

Fig. 10 Dependence of cavity shape from Froude number at zero cavitation number

number and cavitation number the theoretical model gives possibility to build the stationary flows with cavities which are close on horizontal wall and contain a few semi-waves. The cavity shapes which contain $N = 1$, 3 and 5 semi-waves at Froude number 0.6 and zero cavitation number are shown on Fig. 11a. The pressure distribution on wall and wedge cheek for cavities, shown on Fig. 11a, is presented on Fig. 11b.

The pressure distribution on wedge cheek and on the wall in front of wedge is the same at any number of semi-waves. The pressure will be different only on the wall behind cavity. Therefore, the cavity length will be defined by flow character in area of cavity closure on the wall.

In practice, the cavity length can be controlled by position of beginning of solid wall segment behind cavity on the basis of calculation dates. Theoretically, it is possible to define the maximal cavity length at given Froude number for each cavitation number. The cavity length with odd number of semi-waves, when the wave amplitude on last segment is less that preassigned positive number can be

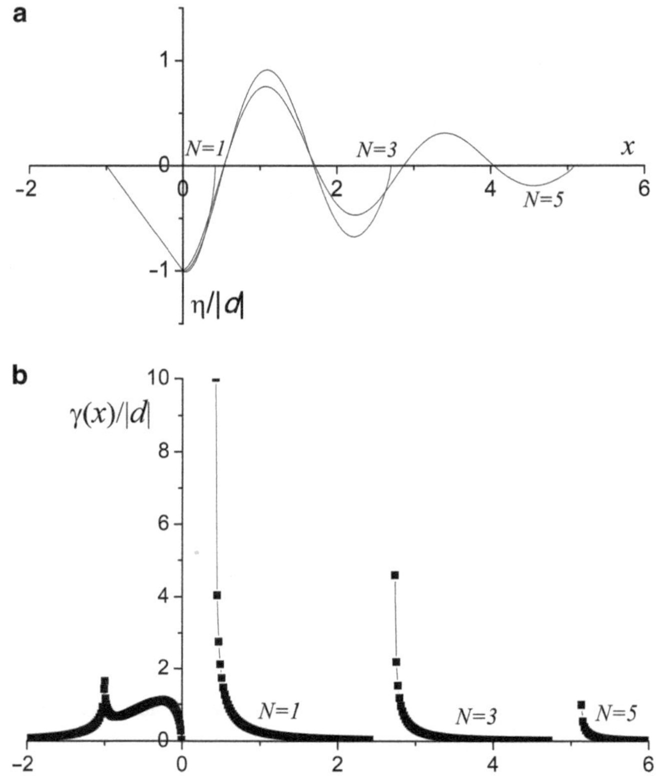

Fig. 11 Cavity shapes (**a**) which consist of different numbers *N*semi-wave and the pressure distribution (**b**)for these cases at $Fr = 0.6$ and $\sigma/|d| = 0$

such characteristic. Note that in previous researches from cavitation behind wedge under solid wall [9, 10], there is no information about decrease of wave amplitude on cavity boundary.

The calculations on the Butuzov's equations and methodic [10] were made for the purpose to make this fact more exact. Let us show the basic results of making calculations. The use of Ryabushinsky's scheme for base and imagine wedges makes the possibility to write the equations system considered in [10]:

$$\frac{1}{\pi}\int_0^1 \frac{q(s)}{x-s}ds - \bar{\nu}\int_0^x q(s)ds + \frac{\sigma}{2} + \frac{\beta}{\pi}\ln\frac{x-1}{x-1-\bar{b}_1} = \bar{\nu}\alpha\bar{c} - \frac{\alpha}{\pi}\ln\frac{x-\bar{c}}{x},\ 0<x<1, \tag{26}$$

$$\int_0^1 q(s)ds - \beta\bar{b}_1 = -\alpha\bar{c}, \tag{27}$$

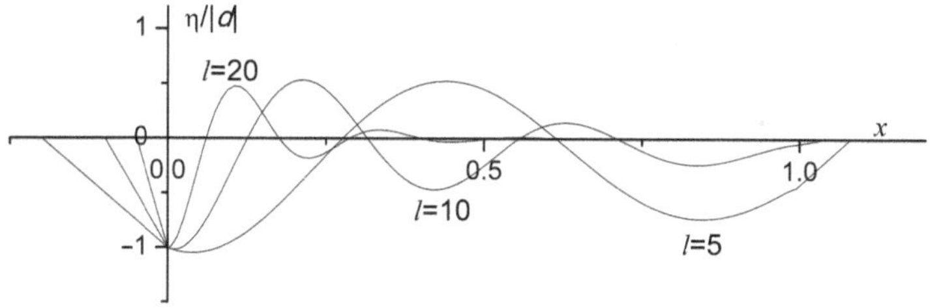

Fig. 12 Modified cavity shapes obtained by Butuzov's method

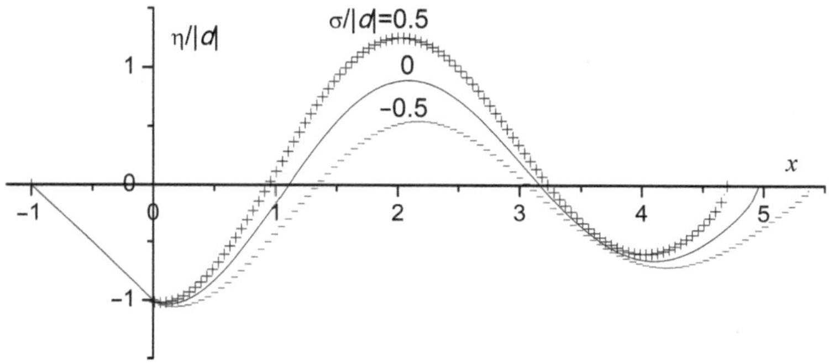

Fig. 13 Cavity shapes at positive, negative and zero cavitation numbers, $Fr = 0.8$

where $\bar{b_1} = b_1/l, \bar{c} = c/l, \bar{v} = vl, b_1$ is the length of imagine wedge, β is the slope angle of imagine wedge, $\alpha$is the slope angle of base wedge. Where the function $q(x)$ does not have singularities and solution of (26) is found in class of functions which are finite on ends of integration interval. The unknowns are cavitation number $\sigma$and the slope angle of imagine wedge β in addition to function $q(x)$. The cavity length l, the length of base wedge c and the length of imagine wedge b_1 are defined.

The cavity shapes which was obtained from solution of (26), (27) at given lengths $l = 5$, 10, 20 and Froude number $Fr = 0.8$ are presented in Fig. 12. As we can see it is also the decrease of wave amplitude in solution on scheme Ryabushinski. The effect of decrease of wave amplitude is more brightly expressed at bigger cavity lengths.

It is possible to estimate the difference in cavity shapes at positive and negative cavitation numbers on data on Fig. 13. The cavities at Froude number 0.8 and cavitation numbers 0, −0.5 and 0.5 are presented there. The cavity lengths correspond to three semi-waves. The cavity shape with positive cavitation number 0.5 is plotted with the help of marker "+", negative with help of "−", zero cavitation number with help of solid line. The characters of cavity curvature and cavity length are changed at negative cavitation numbers.

The curvature can change on opposite with the increase of cavitation number in negative side, i.e. with the increase of pressure in cavity. In this time it is not succeed to satisfy the condition of cavity closure in calculations. For more correct describing of flows at negative cavitaton numbers, it is useful to add the accounting of interaction of fluid and gas flow in cavity into the theoretical model.

In practical problems, there can be requirements of accounting the wall or the depth of the deep under physical conditions. With help of proposed method, it is possible to model the necessary flow conditions by way of matching of appropriated positive or negative cavitation number and geometry of boundaries in area of cavity closure.

Conclusions

The presented method of modeling cavity flows on bottom of planing and displacement hulls gives possibility to resolve the problem in real physical statement and determinate the cavity shape and length at given cavitation number and Froude number.

The wetted lengths of planing surfaces, trim angle and draft are determined in the case of planing hull.

The obtained new results have shown the efficiency of approach and allowed determining the qualitative feature of planing with cavity for real conditions with given displacement and free trim angle.

It is shown that the proposed theory gives the possibility to estimate the ability of natural and artificial cavitation for control to hydrodynamic characteristics of ships.

An example of such estimation is the ability to receive the minimal wave wake and to reduce the wave resistance. This ability can be achieved by way of creation of necessary technical conditions with the help of gas pumping up and appropriated steps constructions.

The research of cavitaton behind wedge under solid horizontal wall with taking into account of gravity forces has allowed determining the wave shape of cavity boundary with decreasing amplitude.

The existence and physical meaning of countable number of cavity lengths is shown. The definition of maximal and minimal cavity length is given.

The tendency of cavity length increasing and bending change at negative cavitation numbers is shown.

The obtained qualitative and numerical results can be used for design and research of air cavity ships for the purpose to reduce the drag and wave resistance.

References

1. Voytkunski YI, editor. The manoeuvrability of displacement ships. Hydrodynamics of ships with dynamic principles of support. Handbook on ship theory. In three volumes. Vol. 3. Leningrad: Sudostroenie; 1985. 544p. (In Russian).
2. Pashin VM, Ivanov AN, Kaliuzhny VG, Lyakhovitsky AG, Pavlov GA. Hydrodynamic design of artificially-ventilated ships. International Symposium on Ship Propulsion dedicated

to the 95-th Anniversary of Professor V.M.Lavrentiev. Proceedings; 19–21 Iune 2001. St. Petersburg; 2001. pp. 117–23.
3. Rusetsky Alexander A. Engineering application of separated cavitation flows in shipbuilding. High speed hydrodynamics. Proceedings of International Summer Scientific School; June 16–23, 2002. Cheboksary. Cheboksary/Washington, DC: Comp. Public.; 2002. p. 93–7.
4. Pavlenko GE. Selected transactions. Kyiv: Naukova dumka; 1979 (In Russian).
5. Knapp RT, Daily JW, Hammitt FG. Cavitation. Moscow: Mir Publishers; 1974 (In Russian).
6. Logvinovich GV. Hydrodynamics of flows with free boundaries. Kiev: Naukova dumka; 1969. In Russian.
7. Savchenko YN. Supercavitation – problems and perspectives. Proceedings of the Fourth International Symposium on Cavitation. California Institute of Technology, Pasadena; 2001
8. Savchenko YN. The research of supercavitation flows. Appl Hydromech. 2007;9(2–3):150–58 (In Russian).
9. Butuzov AA. About limited parameters of artificial cavity which generated on bottom of horizontal wall. Proc Acad Sci USSR Fluid Gas Mech. 1966;2:167–70 (In Russian).
10. Butuzov AA. About artificial cavity flow behind wedge on bottom of horizontal wall. Proc Acad Sci USSR Fluid Gas Mech. 1967;2:83–7 (In Russian).
11. Butuzov AA, Pakusina TV. Solution of flow past a planing surface with an artificial cavity. Trans Acad AN Krylov TsNII. 1973;258:63–81 (In Russian).
12. Barabanov VA, Butuzov AA, Ivanov AN. Detached cavity flow past hydrofoils in the case of planing and in an infinite stream. Non-steady flow of water at high speeds. Proceedings of the IUTAM Symposium Held in Leningrad; June 22–26, 1971. Moscow: Nauka Publishers; 1973. p. 113–9. (In Russian).
13. Rozhdestvenski VV. Cavitation. Leningrad: Sudostroenie; 1977 (In Russian).
14. Ivanov AA. Hydrodynamics of supercavitating flows. Leningrad: Sudostroenie; 1980 (In Russian).
15. Matveev KI. On the limiting parameters of artificial cavitation. Ocean Eng. 2003;30:1179–90.
16. Makasyeyev MV. Stationary planing of a plate over the surface of a ponderable liquid at a specified load and a free trim angle. Appl Hydromech. 2003;5(2, 77):73–5 (In Russian).
17. Dovgiy SA, Makasyeyev MV. Planing of a system of hydrofoils over the surface of a ponderable liquid. Dopovidi NAN Ukrainy. 2003;9:39–45 (In Russian).
18. Makasyeyev MV. Numerical modeling of cavity flow on bottom of a stepped planing hull. Proceedings of the 7th International Symposium on Cavitation (CAV2009); August 17–22, 2009, Ann Arbor; 2009. Paper No. 116. 9p.
19. Vladimirov VS. Equations of mathematical physics. Moscow: Nauka; 1981 (In Russian).
20. Makasyeyev MV. Planing of plate with given load on the surface of heavy fluid. Naukovi visti NTUU “KPI”. 2002;6:133–40 (In Ukrainian).
21. Newmann G. Marine hydrodynamics. Leningrad: Sudostroenie; 1985 (In Russian).
22. Roman VM, Makasyeyev MV. Calculation of the shape of a cavity downstream of a cavitating finite-span hydrofoil. Dynamics of a Continuum with Nonsteady Boundaries. Cheboksary: Chuvashia University Publishers; 1984. p. 103–9. (In Russian).
23. Belotserkovsky CM, Lifanov IK. Numerical methods in singular integral equations. Moscow: Nauka Publishers; 1985 (In Russian).
24. Efremov II. Linearized theory of cavitation flow. Kiev: Naukova dumka; 1974 (In Ukrainian).
25. Himmelblau D. Applied nonlinear programming. Moscow: Mir Publishers; 1975 (In Russian).

Controlled Supercavitation Formed by a Ring Type Wing

Vladislav P. Makhrov

Abstract
The paper presents the some results of theoretical and experimental research of axisymmetric supercavity flow formed by a ring type wing. These flows are known as Lighthill-Shushpanov ones. It has been simulated by distribution of the vortex singularities on combination the "body-ring wing-cavity" surface. Numerical solutions of a set of integral-differential equations were obtained using a spline function for the cavity shape with positive and negative cavitation numbers. The results of the cavitation experimental testing have been cited as an example of the new methods of the cavity formation.

1 Introduction

The idea of supercavitation attracts the attention of creators of high velocity underwater vehicles as a fundamental way to reduce the hydrodynamic drag, and first of all – the friction drag. Consequently, it may increase the vehicle velocity significantly. Problems of organization and calculations of the flows for the cavitation drag decreasing are the main tasks of the supercavitation investigations. A moving system as whole is complicated by different aggregates using for the energy consumption for the drag overcoming and for the gas injection for the cavity ventilation. H. Reichardt [1] obtained the basic tenet and equations for the cavity formation. In spite of many theoretical and experimental investigations of supercavitation, the practical use of cavity flows for underwater motion is rather limited.

It is known that the usual supercavity shape is represented as an ellipsoid. The velocity on its surface is equal to:

V.P. Makhrov (✉)
Moscow Aviation Institute, State Technical University, Moscow, Russia
e-mail: k608@mai.ru

I. Nesteruk (ed.), *Supercavitation*,
DOI 10.1007/978-3-642-23656-3_4, © Springer-Verlag Berlin Heidelberg 2012

$$V_{\sigma} = V_{\infty}(1 + \sigma)^{1/2} = const \quad (1)$$

The body-cavity drag coefficient is proportional to the cavitation number ($C_d \sim \sigma$). It means that for an ordinary supercavity this coefficient can be written as:

$$C_d = C_{do}(1 + \sigma), \quad (2)$$

where C_{do} is the drag coefficient at $\sigma = 0$;

$$\sigma = 2(p_{\infty} - p_{\sigma})/\rho V_{\infty}^2, \quad (3)$$

V_{∞}, and p_{∞} are the free-stream velocity and pressure, p_{σ} – pressure in the cavity, ρ is the density.

The basic formulas had been proposed by Reichardt for the various test conditions. An expanded concept about the current status of the supercavitation research is presented, for example, by V. Serebryakov [2] and E. Paryshev [3] as well. The practical use of these researches was realized first in the high-speed Russian underwater rocket "Shkval" [4, 5]. However, until the present time there is no evidence that the cavity may be controllable yet. The present paper is an overview of investigations of the controlled cavitation flow and its boundary formed by the hydrodynamic singularities.

In 1940s M.J. Lighthill [6] has proposed to use the hydrodynamic singularities for the cavity boundary formation with a negative cavitation number. Later several plane problems about a cavity under the vortex effect have been solved in our country recently. For example, V. Migachev [7] solved such a problem for the pair of vortices. At the Moscow State University (MSU) V. Prokofiev [8] solved a problem for a cavity past a flat plane with the use of the Lighthill's method. In Moscow Aviation Institute (MAI) E. Maraqulin [9] solved the analogous problems using a scheme by the Efros. To confirm theoretical Lighthill's idea in [10] the problem about the horizontal cavitation flow with positive and negative cavitation numbers past a body of revolution formed by an axisymmetric ring vortex was solved.

Systematic physical experiments on cavitation flows formed according to Lighthill method were performed at MSU by Professor Vladimir F. Shushpanov and his colleagues in MAI [11]. Shushpanov showed first that a cavity is formed by the hydrodynamic singularities – the ring wing (annular airfoil) and others hydrodynamic singularities and depends on the geometry of the ring wing, the cavitator and their combinations.

The similar flows were theoretically and experimentally obtained by using the ring type wing [12–14]. We named such cavitation flows as Lighthill-Shushpanov flows. Figure 1 shows the first Lighthill's real cavity. It is obtained by using the practical ring of hydrodynamic singularities – a ring water scoop.

Fig. 1 Cavity by Lighthill with negative cavitation Number (Experiment by Sushpanov)

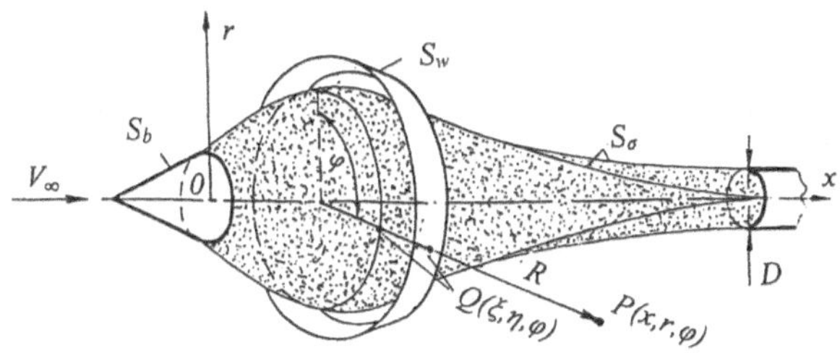

Fig. 2 Scheme of the combination "cavitator – ring wing – cavity"

2 Problem Formulation

2.1 The Bases of Approaches

Imagine the horizontal cavitation flow behind the body of revolution formed in the horizontal potential stream of incompressible and imponderable ideal liquid when the ring wing is under effect of the cavity formation in the unified combination "body – ring wing – cavity". Here, the cavitation number σ may be positive and negative.

Figure 2 illustrates the cavitation flow pattern formed by the ring wing in the cylindrical system of coordinates (x,r,φ). Point Q belongs to the body-cavitator, cavity and ring wing; point P belongs to outward flow accordingly.

Basing on a principle of superposition, the characteristics of this combination are represented by a sum of non-disturbance and disturbance stream functions. It may be written as [15]:

$$\psi = \psi_\infty + \psi_b + \psi_w + \psi_\sigma, \tag{4}$$

where indexes b, w and σ relate to the body-cavitator, ring wing and cavity, respectively.

For the mathematical model composition and problem solution one used a continuous surface of ring singularities – a vortex layer with unknown intensity γ – to form the body, ring wing and cavity as a unitary body. It is known that mathematics of a ring type wing uses combination of the spatial arrangement and the vortex layer method allows the ring wing description with any camber of the hydrofoil to be used as well. However, the hydrofoil contour is limited to be smooth everywhere by Lapunov, excepting the hydrofoil trailing edge. For this $\gamma_w = 0$.

The stream function with radius η of the vortex layer for each component in formula (3) is determined during numeric investigation of this problem:

$$\psi_v = -\frac{r}{4\pi}\int_S \frac{\gamma(\varphi)\cos\varphi}{R}dS = -\frac{r}{4\pi}\int_S \frac{\gamma(\xi,\eta)\eta\cos\varphi d\varphi dl}{R} \tag{5}$$

In (5) $\gamma(\varphi) = \gamma(\xi,\eta)$ is the unknown density function of the vortex layer singularities; $ds = \eta\, d\varphi\, dl$; R is the radius of the vortex layer.

2.2 Characteristics of the Symmetrical Flow Around the Combination of the Body of Revolution and the Ring Wing

The first step of using the vortex layer method is the velocity ratio estimation for the surface of this combination in the non-continuous flow. The main value of the stream function for this case has the form [15]:

$$\psi(x,r) = V_\infty\frac{r^2}{2} - \frac{r}{4\pi}\left[\iint_{S_b}\gamma\frac{\cos\varphi}{R_b}dS + \iint_{S_w}\gamma\frac{\cos\varphi}{R_w}dS\right], \tag{6}$$

where $R_b\ (P,Q_b)$, $R_w\ (P,Q_w)$ are the distances from point $P(x,r)$ in the flow zone to points at the surfaces of the body Q_b and ring wing Q_w.

Equation 4 in terms of Eq. 5 can be solved for the combination of bodies with the use of the boundary conditions:

- The surfaces of body, wing and cavity are impenetrable:

$$\upsilon_n/_S = -(1/r)(d\psi/d\tau)_S = 0;$$

- The stream function on the boundary is constant:

$$\psi/_S = 0;$$

- The change of the tangential velocity is:

$$\upsilon_\tau/_S = (1/r)(d\psi/dn)/_S,$$

where $\bar{n}$ and $\bar{\tau}$ – are the unit vectors to the meridional bodies contour. For combination of the two axisymmetric bodies, the limit values of the normal derivative of the stream function will be a difference on the outside and inside the body and ring wing surfaces.

Considering $R_b(P,Q_b) = R_b(Q_b,Q_{b0}), R_b(P,Q_w) = R_b(Q_{b0},Q_w)$ for body surface, it may be written:

$$\frac{\partial\psi}{\partial n_0}\Big|_e = \gamma_b(Q_0)\frac{\eta_{b0}}{2} - \frac{\eta_{b0}}{4\pi}\Big[\iint\limits_{S_b}\gamma_b\cos\varphi\frac{\partial}{\partial n_b}\left(\frac{1}{R(Q_{0b},Q_b)}\right)\Big|_{Q_{0b}}dS + \iint\limits_{S_w}\gamma_w\cos\varphi\frac{\partial}{\partial n_b}\left(\frac{1}{R(Q_{0b},Q_w)}\right)\Big|_{Q_{0w}}dS\Big] - \psi_{Lb}\frac{\xi'_{b0}(Q_0)}{\eta_{b0}};$$

$$\frac{\partial\psi}{\partial n_0}\Big|_i = -\gamma_b(Q_0)\frac{\eta_{b0}}{2} - \frac{\eta_{b0}}{4\pi}\Big[\iint\limits_{S_b}\gamma_b\cos\varphi\frac{\partial}{\partial n_b}\left(\frac{1}{R(Q_{0b},Q_b)}\right)\Big|_{Q_{0b}}dS + \iint\limits_{S_w}\gamma_w\cos\varphi\frac{\partial}{\partial n_b}\left(\frac{1}{R(Q_{0b},Q_w)}\right)\Big|_{Q_{w0}}dS\Big] - \frac{\xi'_{b0}(Q_0)}{\eta_{b0}}\psi|_{Lb}; \quad (7)$$

where $R(Q_{0b}, Q_w)$ is a distance between point $Q_{0b} \in S_b$ and point $Q_w \in S_w$, indexes e and i are exterior and interior, respectively; index o – is a point on the surface in which the velocity and stream function are determined.

Considering $R_w(P,Q_b) = R_w(Q_{w0},Q_b), R_w(P,Q_w) = R_w(Q_{w0},Q_w)$ for the surface ring wing, the equations for them will be identical

$$\frac{\partial\psi}{\partial n_0}\Big|_e = \frac{\eta_{w0}\gamma_w(Q_0)}{2} - \frac{\eta_{w0}}{4\pi}\Big[\iint\limits_{Sw}\gamma_w\cos\varphi\frac{\partial}{\partial n_w}\left(\frac{1}{R(Q_{0w},Q_w)}\right)\Big|_{Q_{0w}}dS + \iint\limits_{S_b}\gamma_b\cos\varphi\frac{\partial}{\partial n_w}\left(\frac{1}{R(Q_{0w},Q_b)}\right)\Big|_{Q_{0w}}dS\Big] - \frac{\xi'_w(Q_0)}{\eta_{w0}}\psi|_{Lw};$$

$$\frac{\partial\psi}{\partial n_0}\Big|_i = -\frac{\eta_{w0}\gamma_w(Q_0)}{2} - \frac{\eta_{w0}}{4\pi}\Big[\iint\limits_{S_w}\gamma_w\cos\varphi\frac{\partial}{\partial n_w}\left(\frac{1}{R(Q_{w0},Q_w)}\right)\Big|_{Q_{w0}}dS + \iint\limits_{S_b}\gamma_b\cos\varphi\frac{\partial}{\partial n_w}\left(\frac{1}{R(Q_{w0},Q_b)}\right)\Big|_{Q_{w0}}dS\Big] - \frac{\xi'_w(Q_0)}{\eta_{w0}}\psi|_{Lw}; \quad (8)$$

In the system of coordinates the distance between Q_0 and Q is: $R_o = R(Q_o, ;Q) = [(\xi - \xi_o)^2 + \eta^2 + \eta_o^2 - 2\eta\eta_o\cos\varphi]^{1/2}$, where $\eta_o = \eta(\xi_o)$; Q_o $(\xi_o,\eta_o) \in S_a$; Q $(\xi,\eta) \in S$.

If one considers: $V^{(e)}_{\tau 0}(Q_0) = \frac{1}{\eta_0}\cdot\frac{\partial\psi(Q_0)}{\partial n_0}|_e$, $V^{(i)}_{\tau 0}(Q_0) = \frac{1}{\eta_0}\cdot\frac{\partial\psi(Q_0)}{\partial n_0}|_i$, and excepts η_o, then (the internal flow is absent):

$$V^{(e)}_{\tau 0}(Q_0) = \gamma(Q_0).$$

At the condition that vortex intensity is equal to the sum of the tangential velocity in points on the surfaces of the body and ring wing, and there is jump on the vortex layer, the surfaces may be presented by equations for the distribution of velocity $\gamma(Q)$ on the surface for the body and the ring wing in the following form, [16]:

$$\gamma_b(Q_0) = -\frac{1}{2\pi}[\iint\limits_{S_b} \gamma_b \cos\varphi \frac{\partial}{\partial n_b}\left(\frac{1}{R(Q_{b0}, Q_b)}\right)|_{Q_{b0}}\, dS$$
$$+ \iint\limits_{S_w} \gamma_w \cos\varphi \frac{\partial}{\partial n_b}\left(\frac{1}{R(Q_{b0}, Q_b)}\right)|_{Q_{b0}}\, dS] + V_\infty \xi'(Q_{b0}); \quad (9a)$$

$$\gamma_w(Q_0) = -\frac{1}{2\pi}[\iint\limits_{S_w} \gamma_b \cos\varphi \frac{\partial}{\partial n_w}\left(\frac{1}{R(Q_{b0}, Q_b)}\right)|_{Q_w}$$
$$+ \iint\limits_{S_w} \gamma_w \cos\varphi \frac{\partial}{\partial n_w}\left(\frac{1}{R(Q_w, Q_{w0})}\right)|_{Q_{w0}}\, dS$$
$$+ \frac{1}{2\pi}[\iint\limits_{S_b} \gamma_b \cos\varphi \frac{\xi'_w(Q_0)dS}{\eta_{w0}R(Q_b, Q_{b0})}$$
$$- \iint\limits_{S_w} \gamma_w \cos\varphi \frac{\xi'_w(Q_0)dS}{\eta_{w0}R(Q_w, Q_{w0})}] + 2V_\infty \xi'_w(Q_0), \quad (9b)$$

where γ_b and γ_w are identical by the velocity of flows at the surfaces of the body and ring wing, respectively.

Here the integral of S_w in Eq. 9a allows the influence of the ring wing at the flow-around of the central body to be estimated, and analogously integral of S_b in Eq. 9b allows the influence of the central body at the ring wing to be determined. These equations make it possible to calculate the velocity and pressure on the surfaces of this combination. So, this result is applied for further consideration as a base for obtaining the more necessary dependencies in the following calculations of supercavitation flows formed by the ring wing. Hereafter, application of the method of vortex layer allows the ratio of velocities on combination of the body of revolution (cavitator) and the ring type wing to be estimated.

2.3 Characteristics Symmetrical Flow-Around of Combination: "Body – Ring Wing – Cavity"

It is assumed that the cavitation flow formed by the ring wing may correspond to the cavitation figures at $\sigma > 0$ and at $\sigma < 0$; that cavity boundaries closed at the flow axis or on the body surface may be taken in the additional condition. When using Eq. 1 for cavitation velocity and $V_\infty \equiv 1$, interdependencies derivable from (9), we can compose the set of integral-differential equations for the solution of the problem (4). It may be written in the following form:

$$\gamma_b(Q_{b0}) = \xi'_b(Q_{b0}) - \frac{1}{2\pi}[\iint\limits_{S_b} \gamma_b(Q) \cos\varphi \frac{\partial}{\partial n_b}\left(\frac{1}{R(Q_b, Q_{b0})}\right)|_{Q_{b0}} dS$$
$$- \gamma_\sigma \iint\limits_{S_\sigma} \cos\varphi \frac{\partial}{\partial n_\sigma}\left(\frac{1}{R(Q_\sigma, Q_{b0})}\right)|_{Q_{b0}} dS$$
$$- \iint\limits_{S_w} \gamma_w(Q) \cos\varphi \frac{\partial}{\partial n_w}\left(\frac{1}{R(Q_w, Q_{b0})}\right)|_{Q_{b0}} dS]; \tag{10a}$$

$$\gamma_w(Q_{w0}) = -\frac{1}{2\pi}\{\iint\limits_{S_b} \gamma_b(Q) \cos\varphi[\frac{\xi'_w(Q_0)}{\eta_{w0} R(Q_{w0}, Q_b)} - \frac{\partial}{\partial n_{kw}}\left(\frac{1}{R(Q_{b0}, Q_b)}\right)|_{Q_{w0}}] ds$$
$$+ \gamma_\sigma \iint\limits_{S_\sigma} \cos\varphi[\frac{\xi'_w(Q_{w0})}{\eta_{w0} R(Q_{kw}, Q_\sigma)} - \frac{\partial}{\partial n_\sigma}\left(\frac{1}{R(Q_{w0}, Q_\sigma)}\right)|_{Q_{b0}}] dS$$
$$+ \iint\limits_{Sw} \gamma_w(Q) \cos\varphi[\frac{\xi'_{kw}(Q_{w0})}{\eta_{w0} R(Q_{w0}, Q_\sigma)}$$
$$- \frac{\partial}{\partial n_w}\left(\frac{1}{R(Q_{w0}, Q_w)}\right)] dS\} - 2\xi'_w(Q_{w0}); \tag{10b}$$

$$r_\sigma(Q_{\sigma 0}) = \frac{1}{2\pi}[\iint\limits_{S_b} \gamma_b(Q) \cos\varphi \frac{1}{R(Q_{\sigma 0}, Q_b)} dS + \gamma_\sigma \iint\limits_{S_\sigma} \cos\varphi \frac{1}{R(Q_{\sigma 0}, Q_\sigma)} dS$$
$$+ \iint\limits_{S_w} \gamma_w(Q) \cos\varphi \frac{1}{R(Q_w, Q_{\sigma 0})} dS]; \tag{10c}$$

The solution of this set is uniquely determined, if it is complemented by the boundary condition of the cavity closing: $\xi_\sigma (L) = 0$ and $\xi_\sigma'(L) = 1$ – in the case for closing of the boundaries in the point on the axis; $\xi_\sigma (L) = D$ – in the case for closing of the boundaries on the half-infinite cylinder with diameter D – an analog to the scheme by Roshko – Zhukovsky; $\xi_\sigma (L)$ is abscissa of the cavity closing point.

In this problem the diameter D (see Fig. 2) may be unknown, on the surface the velocity changes from $\gamma_\sigma \equiv V_\sigma$ to the velocity of the stream V_∞. In practice the diameter D is given, when the cavity length and the cavitation number are determined. If one should give the cavity length L_σ, then the cavitation number σ for this length should be defined.

The set of Eqs. 10a, 10b, and 10c makes it possible to describe different cavity shapes with positive and negative cavitation numbers. The Eqs. 10a and 10b are solved, and the conditions of impenetrable at the boundaries of the body, ring wing and cavity, and condition $\psi = 0$ on its surfaces are used.

2.4 Characteristics of the Flow Around the Combination "Body – Ring Wing – Cavity" in the Vertical Flow

Figure 3 illustrates the calculation picture of the cavitation flow behind the body of revolution in the falling vertical gravity flow. It is formed by a ring type wing also. The set of equations for the vertical flow should be designed with taking into consideration the gravity influence. It may be applied for the dynamical boundary condition for quasi-steady flow and when $p_\sigma(t) = \text{const}$. Then the velocity of the flow along the axis may be determined by Bernoulli's formula as follows:

$$\gamma_\sigma^b = \sqrt{1 + \sigma_0 \pm \frac{\overline{x}}{Fr^2}} \neq \text{const.}, \tag{11}$$

where: $\overline{x} = \dfrac{\xi_w - \xi_c}{\eta_c}$; $Fr^2 = \dfrac{V_\infty^2}{g\eta_c}$; ξ_c, η_c are the coordinates of the point of the cavity beginning, and the cavitation number σ_o is for these coordinates.

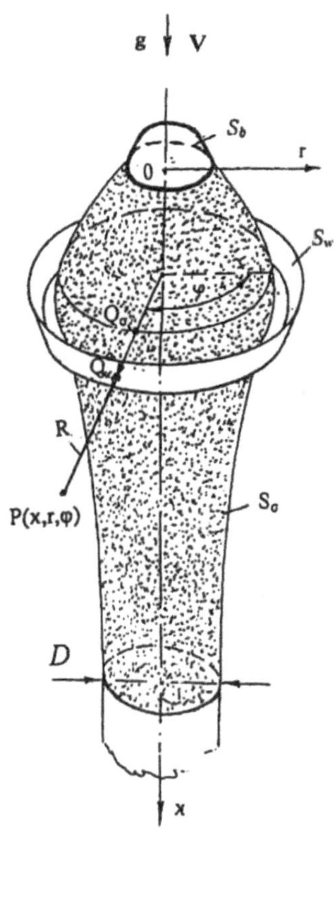

Fig. 3 Scheme of the cavitation flow in the vertical stream

Fig. 4 Estimated cavity boundaries with positive cavitation number: 1 – behind the disk, 2 – behind the cone

The set of equations for this case will be analogous to the set of Eqs. 10a, 10b, and 10c under the suitable conditions of the cavity closing.

3 Numerical Simulation

The existing mathematical model describing the cavitation flow as the set of Eqs. 10a, 10b, and 10c allows all its characteristics to be determined. For the continuous flow it was found that the present method gives good predictions for practical interest. However, numerical difficulties can occur if too much singularities are used to simulate the body and ring wing.

The solution of the set of nonlinear integral-differential equations of the Fredholm's type (10a, 10b, 10c) is obtained by a numerical method. For this purpose, the surface integrals in this set may be converted by a series of integrals identically to the finishing forms has been arrived. Here one can use different approaches, for example, as in [17]. It should be a set of linear algebraic equations, and it is solved by the squaring formula. It is necessary to use a spline fit function also in order to determine the accuracy of the present method for the surface velocity distributions on the body, ring wing and cavity boundaries with exact analytic solutions [18].

Examples of numerical solutions for different schemes of cavities formed by the ring wing using positive and negative cavitation numbers are given here. Details of the method of solving the set of Eqs. 10a, 10b, and 10c are given in [12]. For example, Figs. 4 and 5 illustrate the results of numerical calculations of approximation of real cavity shapes formed past a disk and a cone. Figure 4 shows a cavity with the positive cavitation number (Γ is the relative circulation of the ring wing, $\Gamma = 0.3$). Figure 5 shows a cavity with the negative cavitation number. It illustrates dependencies of the cavity shapes for geometric reciprocal relation body-cavitator and a ring wing.

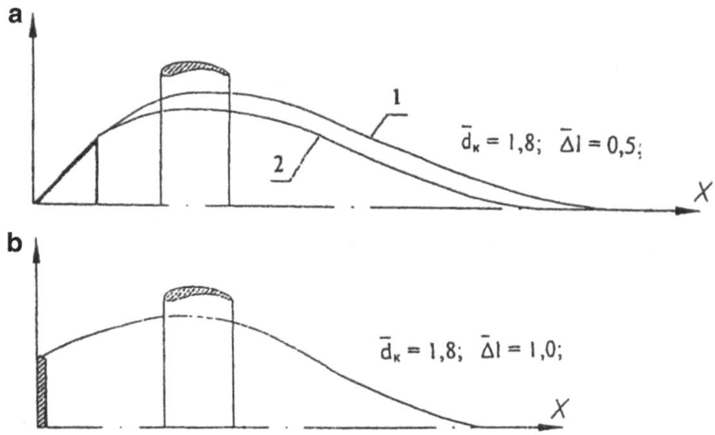

Fig. 5 Estimated cavity boundaries for the cavitation number $\sigma = -0.045$: **a** – behind the cone: 1 – $\sigma = -0.02$, 2 – $\sigma = -0.035$; **b** – behind the disk, $\sigma = -0.045$

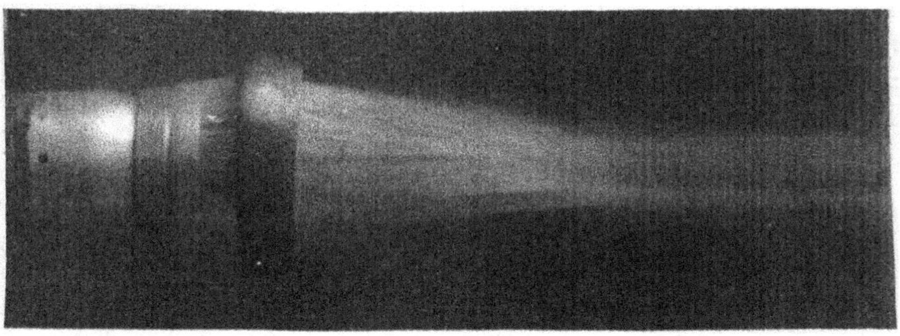

Fig. 6 Tail cavity with a negative cavitation number formed by the ring wing

4 Experimental Data

The theory has been applied to many experimental tests in the hydrodynamic test tunnel, towing and vertical test rig [19]. Experimental procedures were performed for the testing models of bodies of various aspect ratio ($\lambda = 5 - 20$) and for various model's head and the ring wings at all the hydrodynamic test rigs. The cavitation number is simulated by supported gas (air). Measurements of the drag were realized for the qualitative analysis or for the comparison with analogs. The expression for the gas injection rate has the form: $GRT/p_o\, D^2\, V_\infty$. Here, G, R, T are the weight of a

b

c

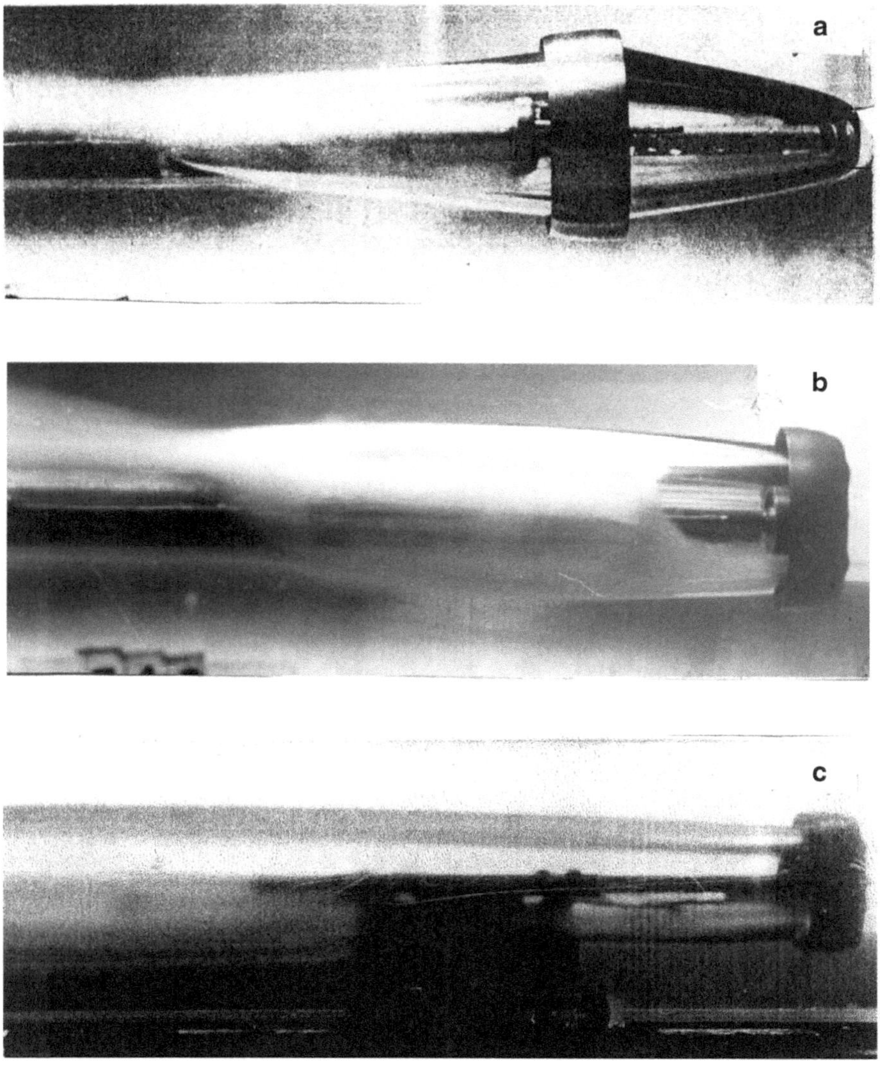

Fig. 7 Examples of a cavity formed by the ring wing: **a**: $\sigma = -0.06$, **b**: $\sigma = 0.12$, **c**: $\sigma = 0.06$

gas injected per 1 s, the gas constant, temperature of gas, respectively, and p_o is the static pressure in the cavitator zone, D is the disk diameter.

Figures 6 and 7 represent several examples of such cavities observed during the laboratory experiments [20]. Experimental data from the water tunnel tests are compared for the well-known data and used in evaluating the present method. Figure 8 shows a comparison of the cavity shapes past the disk without the ring wing (A) with $C_x = 0.89$, and formed by the ring wing around the disk (B) with $C_x = 0.2$. The cavitation number and gas rate C_g were constant here.

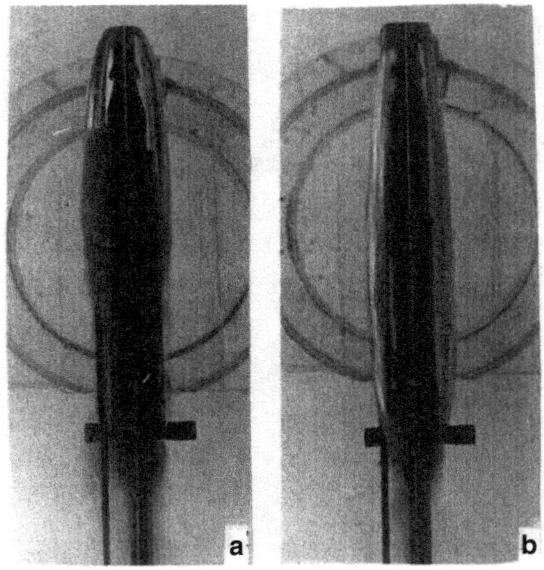

Fig. 8 Example of cavities in the vertical flow ($Fr = 11.2$): **a** – behind the disk, **b** – behind the disk with the ring wing

Conclusions

This paper is the first publication about the new method of forming the controlled supercavitation flow applied earlier by Vladimir Shushpanov.

The most important characteristic of the supercavitation flow formed by the ring wing is the possibility to make the closing flow with the minimal cavitation drag and minimal gas loss from the cavity.

We hope the new method of the supercavitation flow formation stimulates further theoretical and experimental investigations.

References

1. Reichardt H. The laws of cavitation bubbles at axially symmetrical bodies in a flow. Rep. and Translations, N 766. Moscow, Ministry of Aircraft Production; 1946. p. 82.
2. Serebryakov VV. Some problems of hydrodynamics for high speed motion in water with supercavitation. International Conference "Super FAST 2008". St/Petersburg; 2008.
3. Paryshev EV. The dynamics theory of supercavitation. Proceeding of Scientific School «High Speed Hydrodynamics». Cheboksary; 2002. p. 55–70.
4. Sokoliansky VP. Research of high-speed hydrodynamics in hydrodynamics department of CAHI. Proceedings of Conference: The Problems of Body Motion in Liquid with High Velocity. Moscow: CAHI-REGION; 2002. p. 19–30. (In Russian).
5. Shahidzhanov ES. Velocity and high-velocity underwater vehicles. Proceedings of Conference: The Problems of Body Motion in Liquid with High Velocity. Moscow: CAHI-REGION; 2002. p. 3–18 (in Russian).
6. Lighthill MJ. A note on cusped cavities. Aeronautical Research Council. Rep. & Mem. No. 2328.

7. Migachev VI. Axially flow-around wedge in the presence of vortexes. Proceedings of LIEWT, N 113. Leningrad; 1967. p. 34–46. (In Russian).
8. Prokofjev VV. The task about ventilated gas bubble past a plate. Report IM MSU N1550. Moscow; 1974. (In Russian).
9. Marakulin EM. Influence of vortex singularities at the geometric cavity with a recurrent stream. In Set of Papers: Hydromechanics and Energy of Underwater Vehicles. Moscow: MAI; 1991. p. 28–32. (In Russian).
10. Marakulin EM, Makhrov VP, Uzbashev AV. About thin axisymmetric cavity with a isolate ring singularities. In Set of Paper: Hydromechanics and Power engineering of Underwater Vehicles. Moscow: MAI; 1991. p. 23–7 (in Russian).
11. Shushpanov VF, Makhrov VP, Marakulin EM, Kerin NV. The results of experiences for large-scale models with stern-placed cavity formed by a ring type wing in tower tank of A.N. Krylov CNII. Lecture of Conference VMF MO USSR and SUDPROM. Sevastopol; 1984.
12. Makhrov VP. Theoretically investigations of cavity flows with negative numbers. Aerosp MAI J. 2001;8(2):30–9 (In Russian).
13. Makhrov VP, Kerin NV, Pushkarev AA. The using of supercavitation with negative cavitation numbers for high-speed motion under water. Proceedings of Conference: The Problems of Body Motion in Liquid with High Velocity. Moscow: CAHI-REGION; 2002. p. 140–50. (In Russian).
14. Makhrov VP. Hydrodynamic of control flows has shaped free boundaries using of external hydrodynamic singularity – the ring wing. Aerosp MAI J. 2009;16(5):264–73. Moscow: MAI-Print (In Russian).
15. Lumb G. Hydrodynamics. Moscow: Gostechizdat; 1947.
16. Sobolev SL. Equations of Mathematical Physics. Moscow: NAUKA; 1966 (In Russian).
17. Guzevsky LG. Numerical analysis of cavitation flows. Preprint N 40-79 of CO AS USSR. Novosibirsk: Heat-Physics Institute; 1979. (In Russian).
18. Inove M, Kuroumaru M, Jamacuchi S. A solution of Fredholm integral equation by means of the Splaine fit approximation. J Comput Fluids. 1979;7(G.B.):33–46.
19. Kerin NV, Makhrov VP, Pushkarev AA. An experimental research cavity flow with negative cavity number. Proceeding of Conference: The Problems of Body Motion in Liquid with High Velocity. Moscow: CAHI-REGION; 2002. p. 146–150. (In Russian).
20. Grumondz VT, Korzhov DN, Makhrov VP. Some model dynamics problems of high-speed underwater motion of a vehicle with stern-placed wing. International Conference "Super FAST-2008"; July 2–4, 2008. St. Petersburg; 2008.

Drag Effectiveness of Supercavitating Underwater Hulls

Igor Nesteruk

Abstract

The important problem of the drag reduction of underwater hulls was investigated analytically and numerically. The axisymmetric flows of the ideal and the viscous fluid were used. Different effectiveness criteria, such as: the volumetric drag coefficient, the drag coefficients, based on the maximum body cross-section area and the squared hull length, and the ranges of the inertial motion were applied.

The use of known analytic dependences for the slender axisymmetric cavity shapes after the slender or the non-slender cavitators, it was shown that the value of the volumetric drag coefficient and the similar coefficients, based on the squared values of the length and the caliber, can sufficiently be reduced at small cavitation numbers. The smallest values of these drag coefficients correspond to the largest aspect ratios and the slender cavitators. Comparison of the drags of the supercavitating and unseparated flow patterns showed the existence of the critical values of the volume and dimensions. The supercavitating flow pattern is preferable for the values of these parameters smaller than critical ones. The need of the buoyancy force compensation sufficiently diminishes the critical values of the vehicle volume or its dimensions, which achieve maximum at a certain value of the motion velocity. In the case of the base cavity existence, the estimations of the supercavitating hull pressure drag and the comparison with the unseparated flow pattern are presented. The critical values of the body volume have a maximum at a certain value of the movement velocity and drastically increase with the aspect ratio increasing.

Maximum range problems are considered for the supercavitating motion of the axisymmetric body on inertia under an arbitrary angle to horizon. Different isoperimetric problems were formulated and solved with the fixed values of the

I. Nesteruk (✉)
Institute of Hydromechanics of National Academy of Sciences of Ukraine, Kyiv, Ukraine
e-mail: inesteruk@yahoo.com

I. Nesteruk (ed.), *Supercavitation*,
DOI 10.1007/978-3-642-23656-3_5,

body mass, kinetic energy, aspect ratio and caliber. Analytic and numeric solutions for the maximal range and the optimal body shapes are obtained. It was shown that infinite small exceeding some critical value of the initial depth can cause a jump of the range and coming to the water surface. The corresponding values of the critical initial depth are calculated.

1 Introduction

The drag reduction of the high-speed underwater hulls is the important and difficult problem due to the very high water density. Really, for two vehicles with the same shape, volume V and speed U moving in water and air respectively the drag can be expressed as follows:

$$X = 0.5C_V \rho U^2 V^{2/3}.$$

The volumetric drag coefficient C_V can be the same for these two vehicles, if the Mach and Reynolds numbers are similar. Then, due to the huge difference in densities:

$$\frac{\rho_{water}}{\rho_{air}} \approx 800,$$

the drag in water can be estimated to be 800 times greater. Therefore, the drag of an underwater vehicle can be reduced by decreasing the area wetted by water, i.e., by changing the unseparated flow pattern (**a**) by supercavitating one (**b**) (see Fig. 1).

In the case of supercavitation the main part of the hull is located inside the cavity (see Fig. 1b), therefore the skin-friction drag can be reduced sufficiently. This idea was developed in many theoretical, numerical and experimental investigations in a lot of countries. The Ukrainian scientific school, leaded by G. V. Logvinovych and Yu. N. Savchenko, contributed sufficiently both in the experimental research of physical principles of supercavitation, theoretical and numerical simulation of this phenomenon, and in the practical applications (see, for example, [1–24]). Important contributions to the experimental investigations of supercavitation have been done with the use of High-Speed Multi-Purpose Water Tunnel available in the Institute of Hydromechanics of National Academy of Sciences of Ukraine. The experiments with supersonic underwater projectiles are a significant achievement of the scientific school, conducted by Yu. N. Savchenko [8]. In Kyiv Institute of Hydromechanics the computer code was developed to calculate the non-steady supercavity flows, vehicle dynamics and stability [9, 10], which was successfully applied in many countries.

The supercavitating flow pattern shown in Fig. 1b yields a large pressure drag, because of the high pressure acting on the cavitator (a part of the hull, wetted by water). In the case of unseparated flow (see Fig. 1a) the pressure drag is near to zero

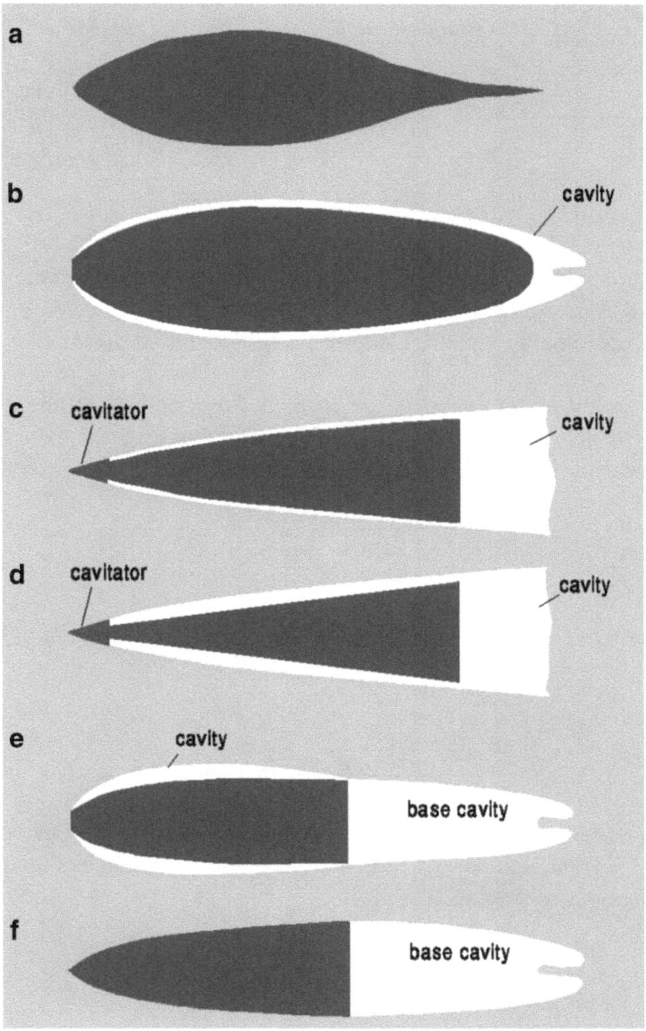

Fig. 1 Different axisymmetric flow patterns

due to the d'Alembert paradox. It was necessary to compare the pressure and skin-friction parts of the total drag and to conclude when the supercavitation is preferable. Such attempts are presented in [25–27]. In this paper the results of these investigations are surveyed and some important conclusions are drowning out.

The flow patterns, shown in Fig. 1a,f correspond to the flow without boundary layer separation and low pressure drag. The supercavitating flow patterns, shown in Fig. 1b–e, ensure low skin-friction drag due to the small surface of the cavitator wetted by water, but the pressure drag can be rather high. To create a cavity, the slender (Fig. 1c,d) and the non-slender (Fig. 1b,e) cavitators can be used. The non-standard flow pattern with a cavity which closes without any artificial closing body

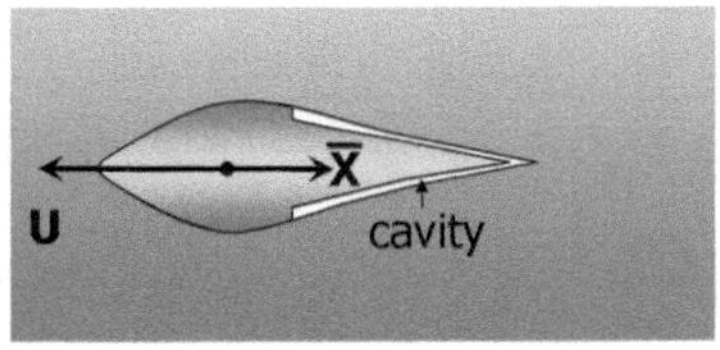

Fig. 2 Non-standard cavitator and cavity which needs no closing body

or re-entrant jet (shown in Fig.2, see also [28]) could provide minimal pressure drag (due to the d'Alembert paradox). The skin-friction drag is reduced in comparison with the unseparated flow pattern shown in Fig. 1a (due to the smaller area wetted by water).

To compare the effectiveness of the different flow patterns, different criteria can be used. If the vehicle velocity U_∞ and the hull volume V_b are fixed the simplest and effective criterion is the volumetric drag coefficient:

$$C_V \equiv \frac{2X}{\rho U_\infty^2 (V_b)^{2/3}} \tag{1}$$

When the hull caliber D_b or its length L_b are fixed, the coefficients C_D or C_L can be used:

$$C_D \equiv \frac{8X}{\rho U_\infty^2 \pi D_b^2}, \quad C_L \equiv \frac{2X}{\rho U_\infty^2 L_b^2} \tag{2}$$

The estimations of C_V for a slender body of revolution without a boundary layer separation are presented in [25, 26]. For the pure turbulent boundary layer the following formula was obtained

$$C_{VU} \approx \frac{0.062}{\lambda_b^{10/21}\,\mathrm{Re}_V^{1/7}},$$

$$\mathrm{Re}_V = \frac{U_\infty {V_b}^{1/3}}{\nu}, \quad \lambda_b = \frac{L_b}{D_b} \tag{3}$$

The C_V estimations for the supercavitating hull which use the total cavity volume (Fig. 1b) can be found in [26] both for slender and non-slender cavitators. In particular, in [27] the following formula was obtained

$$C_V = \sqrt[3]{\frac{9\pi\sigma^4}{-16\ln\sigma}} \tag{4}$$

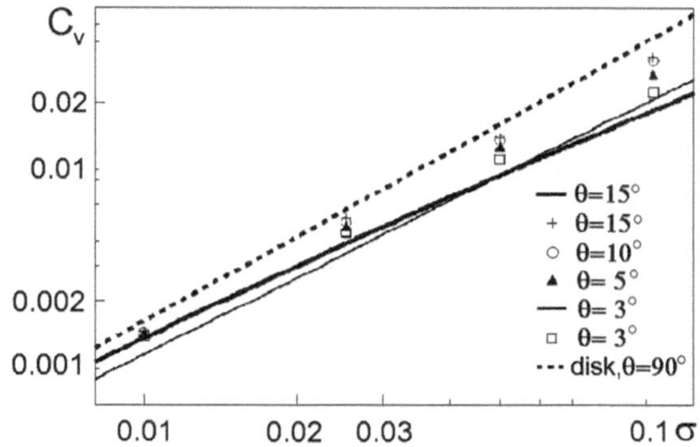

Fig. 3 Volumetric drag coefficients for cones

for conic cavitators with the angle 2θ, $\theta > 25^0$. Equation 4 follows from the well known semi-empirical formulas of Garabedian [29]

$$R^2 = \frac{x(1-x)}{\lambda^2}, \qquad \frac{R_n}{L} = \frac{\sigma}{2\sqrt{-C_x \ln \sigma}},$$
$$\lambda = \frac{L}{D} = \sqrt{\frac{-\ln \sigma}{\sigma}}, \qquad \frac{D}{R_n} = 2\sqrt{\frac{C_x}{\sigma}} \tag{5}$$

Here σ is the cavitation number; $R(x)$ is the cavity radius; R_n is the cavitator radius; λ is the cavity aspect ratio; D is the maximal cavity diameter; L is the cavity length; C_x is the cavitation drag coefficient related to the base section area of the cavitator πR_n^2. It must be noted that the value C_V does not depend on θ for these non-slender cavitators and tends to zero with diminishing of the cavitation number σ. The relationship (5) is represented in Fig. 3 by the dashed line. The results of non-linear numerical calculations for slender cones with the use of the method from paper [30] are presented by dots. The linear calculations with the use of the following formulas (see [15, 17])

$$\frac{R^2}{R_n^2} = \frac{\sigma x^2}{2R_n^2 \ln \beta} + 2\beta \frac{x}{R_n} + 1, \tag{6}$$

$$C_x \approx C_{x0} = -2\beta^2[\ln(0.5\beta) + 1]$$

(β is the derivative of the radius at the point of cavity origin) are shown in Fig. 3 by solid lines.

Unfortunately, for the hull, which uses the total cavity volume (see Fig. 1b), the cavitation number cannot be diminished to zero, since the appropriate cavity aspect ratio λ tends to infinity (see, for example, (5)) when $\sigma \to 0$. The same value of λ_b

has also the hull located in the cavity. The constructive considerations restrict the body aspect ratio. For example, if λ_b is limited by the value $\lambda_m = 20$, the possible cavitation numbers cannot be less than 0.01 for both the slender and the non-slender cavitators, and $C_V \geq 1.5 \cdot 10^{-3}$ (see Fig. 3).

Formula (3) shows that $C_{VU}<1.5 \cdot 10^{-3}$ for $\mathrm{Re}_V>10^7$ and $\lambda_b = 20$. Thus, the standard supercavitating flow pattern (Fig. 1b) is preferable for smaller values of the volumetric Reynolds number $\mathrm{Re}_V<10^7$only. The cavitation number has to be close to the minimal possible value $\sigma \approx 0.01$. The obtained critical value of the Reynolds number $\mathrm{Re}_{V*} = 10^7$means that supercavitation is preferable for very small vehicles. For example, the critical volume of a hull must be limited by values $V^* \approx 10^{-3} - 10^{-6}\mathrm{m}^3$ for the velocity range $100<U_\infty<1000$ m/s. To have a greater effective supercavitating vehicle, we need to diminish the cavitation number. There are two possible ways to realize such decreasing: (1) to use the flow pattern 1b and to increase the body aspect ratio λ_b; (2) to use the initial part of the cavity only (the flow patterns shown in Fig. 1c,d) with the limited values of the hull aspect ratio.

In the first case the critical value of the volumetric Reynolds number can be increased for the hulls with the greater aspect ratio and corresponding less values of the cavitation number. For example, if $\lambda_b = 100$, the corresponding values of the cavitation number and C_V(according to the formulae (4), (5)) can be estimated as follows: $\sigma \approx 0.00072$, $C_V \approx 4 \cdot 10^{-5}$. It means that the drag is 37 times smaller in comparison with the case $\lambda_b = 20$. The supercavitating flow pattern is preferable for such slender hulls (in comparison with the unseparated one shown in Fig. 1a) at all the values of the subsonic velocities and the vehicle dimensions of practical interest.

The strength of the hull limits the aspect ratio increasing. Therefore, the supercavitation drag diminishing is possible with the use the initial part of the cavity only (as shown in Fig. 1c,d). It looks confusing. Really, if the hull uses only a part of the cavity, then the volumetric drag coefficient increases (see formula (1)). But this fact enables us to use smaller cavitation numbers and larger cavities. As a result, the volumetric drag coefficient can be smaller for the flow patterns 1c and 1d in comparison with the 1b one. The detailed proof can be found in [31]. The main results for different values of the maximum hull aspect ratio λ_m are presented in Sect. 2. The problem of the buoyancy compensation for the horizontal movement of supercavitating vehicle is discussed in Sect. 3. The case of the flow patterns with the base cavity (see Fig. 1e,f) is presented (see also [31]). The principal results are reported in Sect. 4.

The maximum range problems are considered in [32] for the supercavitating motion of the axisymmetric body on inertia under an arbitrary angle to horizon in the case of very high velocities and non-slender cavitators. The vehicle can use the initial part of the cavity only. The main results are presented in Sect. 5 for different values of the maximum hull aspect ratio λ_m.

2 The Underwater Hulls Drag Diminishing at Very Small Cavitation Numbers

If the hull is located in the initial part of the cavity only (such as shown in Fig. 1c,d), the appropriate volumetric drag coefficients can be easily defined with the use of (1) and (5) for non-slender cavitator (or (6) in the case of the slender one). The analytical formulas can be found in [31], the calculation examples are presented in Fig. 4 for different values of the maximum hull aspect ratio λ_m. The lines correspond to the non-slender cavitators (the results do not depend on θ); the dots show the case of the slender cavitator with $\beta = 0.1$.

It can be seen from Figs. 3 and 4 that C_V can be sufficiently reduced for $\sigma<0.01$. The smallest values of C_V correspond to the largest values of the hull aspect ratio. In the case of the non-slender cavitators the function $C_V(\sigma)$ has a minimum. May be this fact is connected with the limited accuracy of the Garabedian formulae (5) for very small cavitation number in the region close to the cavitator. Usually, the slender cavitators yield smaller values of C_V. May be it is due to the limited accuracy of the Eq. 6. In any case this interesting fact needs additional investigations with the use of the second approximation equation [18] or a nonlinear approach.

An example of the optimal shape with $\lambda_m = 10$, $\beta = 0.1$, the velocity 700 m/s (the corresponding value of the cavitation number at small depth without ventilation is 0.0004), $L_b / R_n = 62.62$, $C_V = 0.00099$ is shown schematically in Fig. 1c. The optimal hull shape must be as close as possible to the initial part of the cavity. If the hull caliber or its length is fixed, the optimal hull must be only inscribed into the initial part of the cavity, but its caliber must coincide with the diameter of the cavity at the body end (see Fig. 1d). The dependences for coefficients C_D and C_L, which can be obtained with the use of formulae (2), (5) and (6), are shown in Figs. 5 and 6

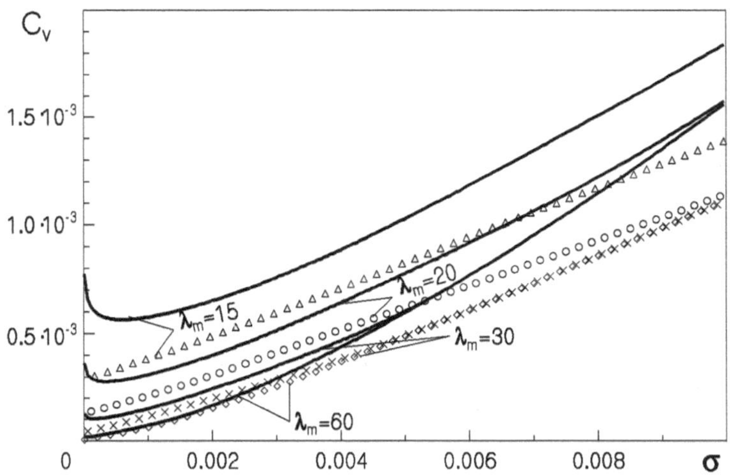

Fig. 4 Volumetric drag coefficients for different values of the hull aspect ratio

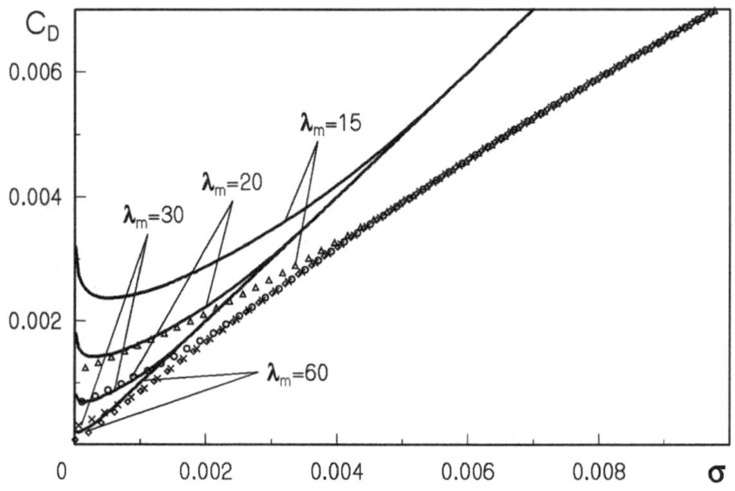

Fig. 5 The drag coefficients C_D for different values of the hull aspect ratio

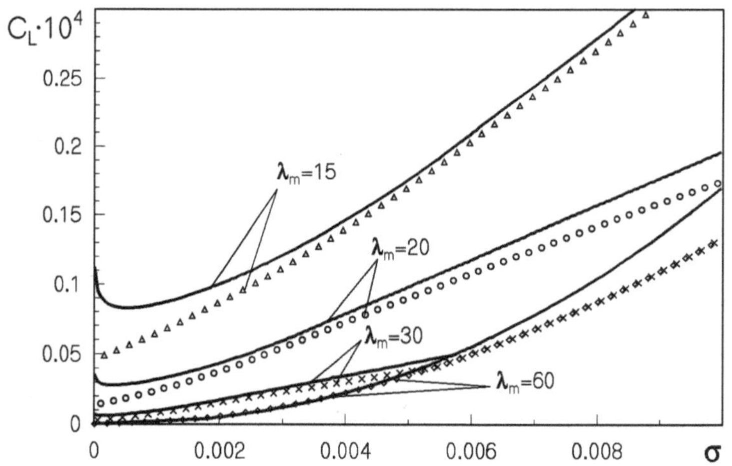

Fig. 6 The drag coefficients C_L for different values of the hull aspect ratio

(see details in [31]). The lines correspond to the non-slender cavitators (the results do not depend on θ); the dots show the case of the slender cavitator with $\beta = 0.1$.

The following equation

$$C_V = C_{VU} \tag{7}$$

can be used to calculate the critical value of the volumetric Reynolds number Re_{V*} which corresponds to the equal efficiency of the unseparated (Fig. 1a) and supercavitating (Fig. 1b, c, d) flow patterns. The supercavitating hull is preferable for $Re_V < Re_{V*}$. With the use of (3) and (7) the following formula can be obtained:

$$Re_{V*} \approx \frac{3.5 \cdot 10^{-9}}{\lambda_m{}^{10/3} C_V^7} \tag{7a}$$

The drastic diminishing of C_V showed in Fig. 4 enables to increase the value of Re_{V*} at small cavitation numbers without increasing the hull aspect ratio. For example, the optimal shape with $\lambda_m = 20$, the non-slender cavitator, the velocity 1,000 m/s and a small depth of the horizontal movement (the corresponding value of the cavitation number is 0.0002, $C_V = 0.00028$) yields the critical values $Re_{V*} \approx 1.2 \cdot 10^{12}$, $V_* \approx 10^9 \text{m}^3$. Therefore, the supercavitating flow pattern is preferable for all possible vehicles of practical interest. It must be noted that supercavitation is preferable even at ten times greater cavitation number (smaller velocity or greater depth of movement). For example, at $\sigma = 0.002$ the critical Reynolds number can be estimated as $Re_{V*} \approx 10^{11}$. But for $\sigma = 0.004$ the critical value decreases to be $Re_{V*} \approx 3.6 \cdot 10^9$and some dimension limitation are possible for the effective supercavitating vehicle. For example, at the velocity 300 m/s a supercavitating hull of volume1 m^3, designed for $\sigma = 0.004$ and $\lambda_m = 20$, has the value of drag approximately 1.5 times smaller than an unseparated vehicle of the same volume and velocity. But for the volume 8000 m^3 the unseparated flow pattern is preferable.

These optimistic predictions of the supercavitating hulls efficiency must be analyzed by taking into account several important things. First of all, the very small cavitation numbers, which are necessary to reduce the drag, are possible at very high velocities only. Otherwise very intensive ventilation is necessary. The large ventilation rates need an additional energy supply which diminishes the efficiency of supercavitation. If we use only vapor cavities, the depth h of movement is limited by the simple equation:

$$h = \frac{\sigma U_\infty^2}{2g} - \frac{p_a}{\rho g}, \tag{7b}$$

where p_a is atmospheric pressure, g is gravity acceleration.

Formulae (7a) and (7b) can be used to calculate the dependencies of the critical volume and critical velocity $U_{\infty *}$ versus the depth of movement h. The examples of calculations for a hull with $\lambda_m = 20$ are presented in Fig. 7a,b. Solid lines correspond to a non-slender cavitator, dashed ones represent a case of the slender conical cavitator with $\beta = 0.1$. Natural supercavitation is preferable in domains located below the corresponding lines in Fig. 7a.

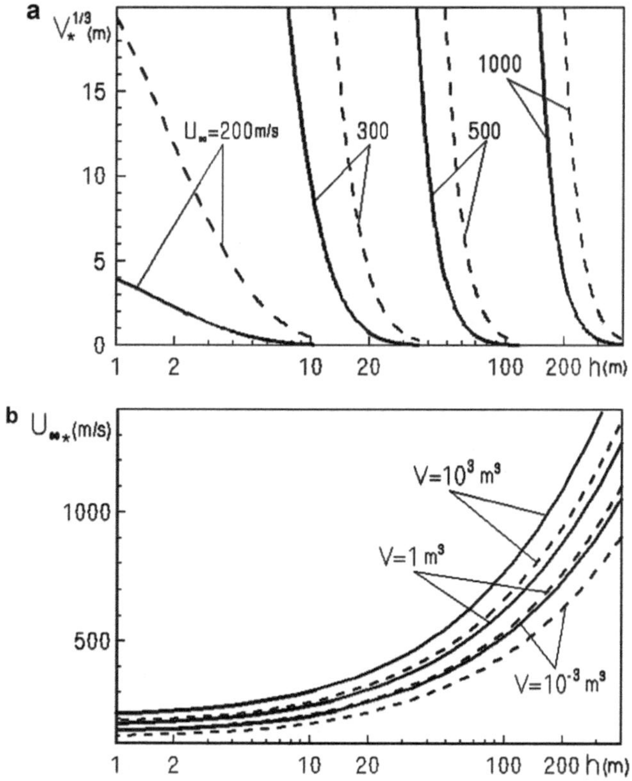

Fig. 7 (**a**) The critical volume of the effective supercavitating hull for different values of the velocity. (**b**) The critical velocity of the effective supercavitating hull for different values of the volume

It can be seen that to achieve a large depth, dimensions of an effective supercavitating vehicle have to be diminished and its velocity to be increased. For a large vehicle at moderate velocity an unseparated flow pattern is preferable to increase the depth.

It must be noted that the slender cavitators can ensure a smaller volumetric drag coefficient (see Fig. 4). It means that the critical volume can be increased (see Fig. 7a). Due to the nonlinear dependence (7a) this increase may be rather great. For example, if a slender cavitator yields a 30% smaller value of C_V, then the critical volume of an optimal hull is approximately 1,790 times greater in comparison with the case of a non-slender cavitator. Therefore, further theoretical and experimental investigations of the optimal hulls with slender cavitators are very important to answer the questions: do really the slender cavitators yield smaller values of C_V and are these cavitators applicable for the high-speed underwater vehicles?

3 The Buoyancy Force Compensation for Supercavitating Vehicles

The supercavitating hull moves in the gas (see Fig. 1b,c,d) with a very small value of the buoyancy force in comparison with the case of the vehicle wetted by water shown in Fig. 1a. Therefore, for supercavitating vehicles, which move horizontally, the problem of their weight compensation must be solved (as in the case of airplanes). For this purpose the hull planing on the cavity surface or underwater wings are used. This situation causes an additional drag ΔC_V, which can be estimated with the use of the aerodynamic effectiveness $k = C_y/C_x$. To calculate the critical Reynolds number, a new equation

$$C_V + \Delta C_V = C_{VU} \tag{8}$$

should be solved instead of (7).

The numerical examples are presented in Figs. 8 and 9. The values $\nu = 1.3 \cdot 10^{-6}\mathrm{m}^2/\mathrm{s}$, $k = 10$ (solid lines) and $k = 1$(dots) were used for calculations with the use of (8).

From Fig. 8 it can be seen that dependencies have a maximum. The presented in [31] analysis enables obtaining the maximum values of the critical volume V_{*m} and the velocity corresponding to this maximum (see details in [31]). Thus, at given volume of a vehicle it is possible to have no effective supercavitating flow pattern at any velocity, if $V > V_{*m}$. It must be noted that in the previous case (without buoyancy force compensation) for every dimension of the hull, a critical velocity $U_{\infty *}$ existed, and the supercavitation was preferable for $U_\infty < U_{\infty *}$ (see Eq. 7a). The influence of the buoyancy force causes the existence of an effective supercavitating hull at $V < V_{*m}$ only in the velocity range $U_{*1} < U_\infty < U_{*2}$ (see Fig. 8). The preferable

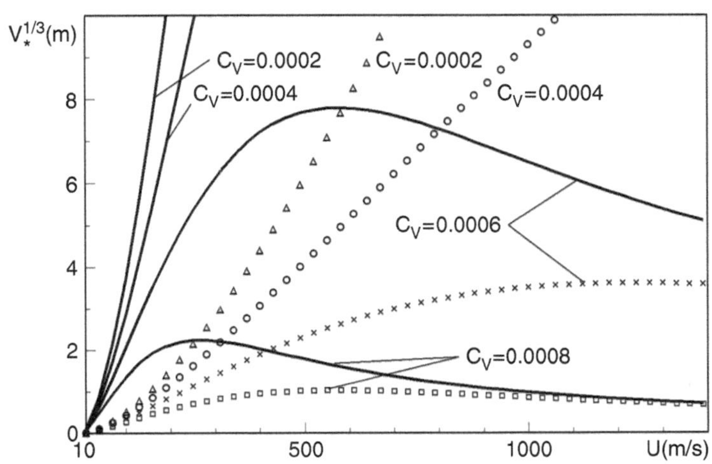

Fig. 8 Dependencies of the critical volume at different values of the volumetric drag coefficient

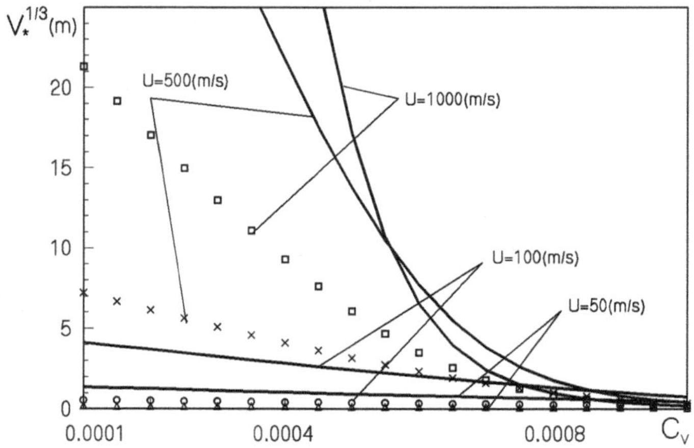

Fig. 9 Dependencies of the critical volume at different values of the velocity

velocity range increases when the value of C_V decreases, in particular U_{*2} can exceed the velocity of sound in water.

Figure 8 shows that the critical volume decreases drastically with the increase of C_V. It means that supercavitation is effective at very small cavitation numbers and for not very great depth (if the vapor cavitation is used). Figures 8 and 9 show also that the necessity of the buoyancy force compensation diminishes the critical volume (especially at small values of k and large values of C_V). The same estimations of the critical hull calibre and its length can be done in the cases when these parameters are fixed (see details in [31]).

4 Comparison of the Supercavitating and Unseparated Flow Patterns with Base Cavities

For the flows with the base cavity there are two options: (1) the hull is covered by another cavity (the two-cavity flow pattern shown in Fig. 1e); (2) the hull is wetted by the water flow without the boundary layer separation as shown in Fig. 1f. The comparison of efficiency of these two patterns was done in [31] with the use of formulae (6) for the pattern 1e and the parabolic unseparated shape

$$\frac{R}{R_n} = \frac{bx^2}{R_n^2} + 1, \quad -\sqrt{-\frac{1}{b}} \le \frac{x}{R_n} \le 0$$

for the flow pattern 1f. Equation 8 was used to calculate the critical volume. The numerical examples are presented in Fig. 10. The values $\nu = 1.3 \cdot 10^{-6}\text{m}^2/\text{s}$, $\beta = 0.1$, $k = 10$ (solid lines) and $k = 1$ (dots) were used for calculations. It can be seen from Fig. 10 that some curves have a maximum (similar as ones shown in

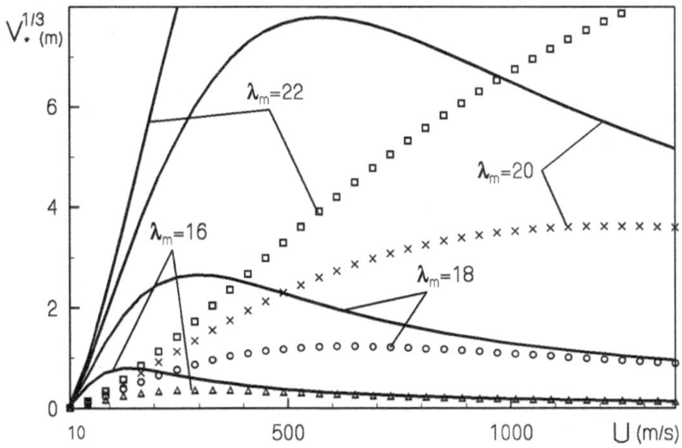

Fig. 10 Dependencies of the critical volume at different values of the hull aspect ratio for the base cavity flow pattern

Fig. 8). The corresponding velocity increases with the hull aspect ratio increasing and may approach to the sonic velocity in water. The critical volume increases drastically with the increasing of the hull aspect ratio. Figure 10 shows also that the buoyancy force compensation diminishes the critical volume (especially at small values of k and large values of λ_m)

5 Optimization Problems for High-Speed Supercavitation Motion on Inertia

The results obtained in the previous sections stimulated the investigation of the effectiveness of the supercavitating flow pattern for the inertial motion with very small cavitation numbers. The horizontal supercavitating motion on inertia and the problem of range maximization were considered by Putilin, Gieseke, Serebriakov, Kirschner, Schnerr [14, 33–35] and other authors. The case of the non-horizontal inertial motion with different isoperimetric conditions was investigated in [26, 36–39], but it was taken into account only the case of complete using the cavity volume (Fig. 1b). The partial cavity use (Fig. 1c,d) is typical for very small cavitation numbers and was investigated in [32] for non-slender cavitators. Here the results of the papers [32, 36–39] will be shortly reported.

Let a model start a rectilinear movement in water on inertia with the velocity U_0 under an arbitrary angle γ to horizon. The distance S, passed by the supercavitating body, should be maximal (see Fig. 11). It was shown (see, e.g., [37, 38]) that in many cases the flow may be supposed to be quasi-stationary and the gravity effect on the cavity and body motion may be neglected. If the cavitator is non-slender, the semi-empirical relations (5) by Garabedian may be used with the current cavitation number σ at the cavitator immersion depth. If we neglect changes of the cavitation

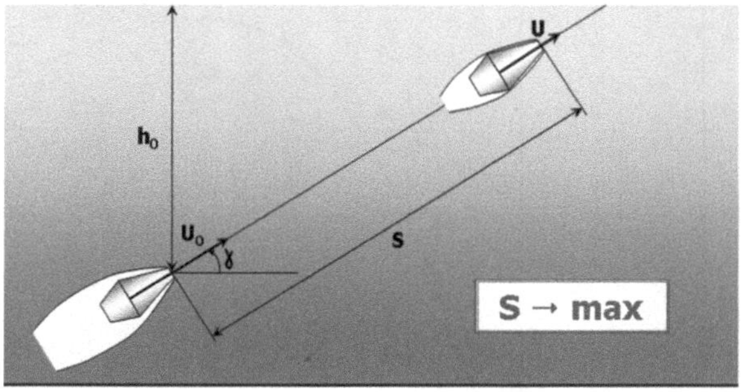

Fig. 11 The maximum range problem for the supercavitating motion on inertia

number $\sigma << 1$, then C_x may be considered to be constant and the distance S passed by the body is defined by the following formula:

$$S = \frac{2m}{\rho C_x \pi R_n^2} \ln \frac{U_0}{U}, \tag{9}$$

where m is the body mass; U is the final body velocity.

In [36–39] Eq. 9 was analyzed for different isoperimetric conditions for the case of complete use of the cavity volume (see Fig. 1b).

5.1 Complete Use of the Cavity Volume

For example, in the case of the fixed starting velocity the most interesting tasks may be listed as follows:

1. The body mass and its caliber D_b are fixed
2. The body mass and its length L_b are fixed
3. The body mass and its volume V_b or the average body density $\rho_b = m/V_b$ and its volume are fixed
4. The average body density and its caliber are fixed
5. The average body density and its length are fixed

Taking into account that a body practically stops after washing off by water, it could be shown that the optimal body shape must coincide with the cavity shape in the moment of washing off for problems 3–5 (see [36] and also [14, 34]). For the isoperimetric conditions 1 and 2, the optimal body shape must be inscribed in the cavity corresponding to the moment of washing off; its caliber must coincide with the cavity caliber for the task 1 and its length must coincide with the cavity length for the task 2. For these two cases the examples of the optimal body shapes are presented in [36]. These five isoperimetric conditions can be applied for the problems with the fixed final depth of supercavitating motion [36, 37] and for the tasks with the fixed initial body depth too [36–38].

Another group of problems arises when the initial (or final) kinetic energy of a body or its initial (or final) momentum are fixed. In these cases the initial velocity and body mass are not fixed. The solution of the problem must answer the question: what are the best values of the initial velocity and body mass to achieve the maximal range. For example, is it better to use a small body with a large velocity or a large body with a small velocity? For the case of horizontal motion the optimum body mass was calculated by Gieseke [33]. For every problem with the non-horizontal body motion, the condition of the fixed initial or final depth can be used. For this group of problems the optimal body shape coincides with the shape of the cavity at the smallest possible cavity number. Usually we use the value $\sigma^* = 0.01$, because it is practically impossible to use shapes with the aspect ratio greater than 20. It is difficult to describe all 18 isoperimetric problems in details. Let us consider their main features and present some examples.

5.1.1 Problems with the Fixed Initial Velocity and the Fixed Final Depth

Such isoperimetric conditions were considered in [36]. In this case all the results (for example, the maximum ranges) do not depend on the angle of motion γ. This conclusion is in good agreement [36] with the calculations, performed with the use of the computer program SCAV [9, 10].

For example, the solution of the fifth problem yields the optimum value of the final velocity (see [36–38])

$$\bar{U}^* = e^{-0.5} \approx 0.607, \tag{10}$$

The maximum range and the optimal value of the final cavitation number are given by the following formulae:

$$\bar{S}^* = \frac{\bar{\rho}_b Fr_0^2}{3e\bar{h}_2}, \quad \sigma^* = \frac{3e\bar{h}_2}{Fr_0^2}, \quad \bar{h}_2 \equiv \frac{10 + h_f}{L},$$

where $\bar{\rho}_b = m/V\rho$; $Fr_0 = U_0/\sqrt{gL}$ is the initial Froude number. Knowing the optimal value of the final cavitation number and the fixed body length (which has to be equal to the cavity length of the moment of washing off, because in this case the optimal body shape coincide with the shape of the cavity at $\sigma = \sigma^*$), the optimal body shape and the optimal cavitator radius can be calculated with the use of formulas (5), see [36–38]. For the isoperimetric problems 1–4, the corresponding values of the optimal dimensionless final velocity $\bar{U}^*$ and the maximum range were obtained in [36].

5.1.2 Problems with the Fixed Initial Velocity and Depth

In this case the cavitation number depends on the motion angle γ

$$\sigma = \frac{2g(h_1 - S\sin\gamma)}{U_0^2\bar{U}^2}$$

where the initial depth h_0 (the initial position of the center of the cavity separation cross-section) and $h_1 = 10 + h_0$ are measured in meters. Therefore, the maximum range and the optimal value of the final cavitation number depend on the motion angle, but the optimum values of the final velocity are the same or very close to those, obtained for the case with the fixed final depth (see [38]). In particular, for the problems 1 and 5 this value coincide with (10), for problems 2–4 it was obtained $\bar{U}^* = e^{-0.25} \approx 0.78$, $\bar{U}^* = e^{-0.375} \approx 0.68$ and $\bar{U}^* = e^{-1/3} \approx 0.72$ respectively. The solution of the fifth problem can be found in [37].

The values of the maximum range have to be calculated with the use of nonlinear equations. For example, for the problems 1 and 5 this equation can be written as follows:

$$\bar{S}^*(\bar{h}_1 - \bar{S}^* \sin\gamma) = \frac{2}{e\pi} \tag{11}$$

The dimensionless values are connected with the physical characteristics by the formulas

$$\bar{S} = \frac{SD}{U_0}\sqrt{\frac{\rho g}{m}},\ \bar{h}_1 = \frac{h_1 D}{U_0}\sqrt{\frac{\rho g}{m}} \text{ or } \bar{S} = \frac{S}{U_0}\sqrt{\frac{6g}{\pi L\bar{\rho}_b}},\ \bar{h}_1 = \frac{h_1}{U_0}\sqrt{\frac{6g}{\pi L\bar{\rho}_b}}$$

Solutions of the Eq. 11 are shown in Fig. 12 by solid lines.

The quadratic equation (11) could be solved analytically. For the problems 2–4 the corresponding equations were solved numerically and gave the similar dependences on the dimensionless initial depth $\bar{h}_1$ and the angle of motion, [38]. Knowing the maximum range, the optimal final cavitation number can be found from the simple relationship obtained in [38] for dimensionless value $\bar{\sigma}^*$:

$$\bar{\sigma}^* = \bar{h}_1 - \bar{S}^* \sin\gamma, \tag{12}$$

where the formulas for $\bar{\sigma}^*$ depend on the isoperimetric conditions. For example, for the first problem this relationship can be written as follows:

$$\bar{\sigma}^* = \frac{\sigma^* D U_0}{2}\sqrt{\frac{\rho}{gm}}$$

The relationship (12) is shown in Fig. 12 by dotted lines. Knowing the optimal value of the final cavitation number, the optimal body shape and the optimal cavitator radius can be calculated with the use of formulas (5).

For $\gamma>0$ the relationships $\bar{S}^*(\bar{h}_1)$ may achieve the infinite slope at some critical values of the initial depth $\bar{h}_1^{(cr)}$ (see Fig. 12). It means that for smaller values of the initial depth the body can reach the free water surface without a loss of the supercavitating flow pattern. The critical values $\bar{h}_1^{(cr)}$ have been found in [38] for

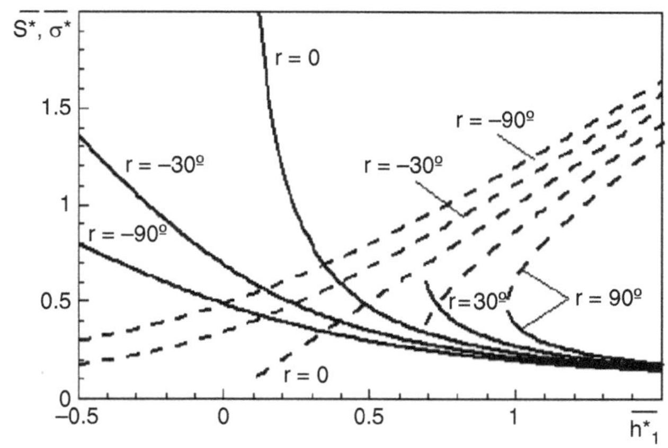

Fig. 12 Dependences of the dimensionless maximum range and the optimal final cavitation number at different angles of motion for the problem 1 or 5

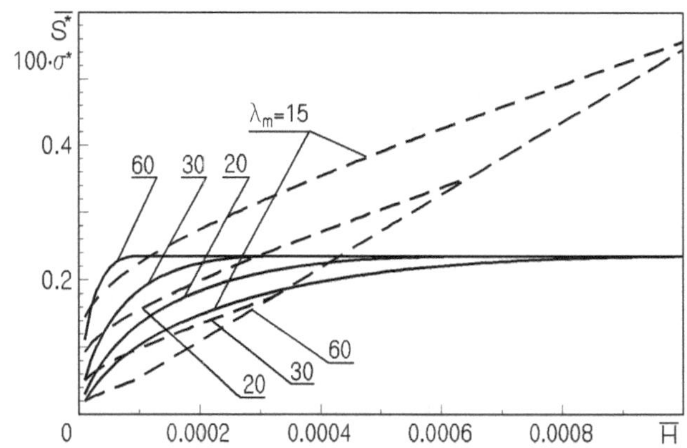

Fig. 13 Dependences of the maximum range and optimal cavitation number for different values of the hull aspect ratio

different isoperimetric conditions. For example, for the problems 1 and 5 the following formula was proposed:

$$\bar{h}_1^{(cr)} = \sqrt{\frac{8 \sin\gamma}{e\pi}} \tag{13}$$

This equation coincides with the corresponding relationship from [37].

5.1.3 Problems with the Fixed Staring or Final Momentum

For the fixed starting momentum $I_0 = mU_0$ and the fixed final depth, the following formulae have been obtained, [37]:

$$\bar{U}^* = e^{-3} \approx 0.0498,\ S^* = \frac{6a\sigma^{1/6}}{eC_V},\ a = \frac{\rho_b^{2/3} I_0^{1/3}}{\rho\left[2g(h_f + 10)\right]^{1/6}}.$$

For all four isoperimetric problems the optimal value of the optimal final cavitation number σ^* must be as small as possible and the optimal hull shape must coincide with the cavity shape at this cavitation number. For example, when $\sigma^* = 0.01$, then $S^* \approx 654a$. The optimal values of the body mass can be obtained for the cases with the fixed starting and final momentum respectively:

$$m^* = \frac{0.00352 I_0}{\sqrt{g(h_f + 10)}},\ m^* = \frac{0.071 I_f}{\sqrt{g(h_f + 10)}}.$$

Then the optimal body volume, length, caliber and cavitator radius can be calculated with the use of Eq. 5. To determine the maximum range in the problems with the fixed starting depth, it is necessary to solve the non-linear equations, similar to (11).

5.1.4 Problems with the Fixed Staring or Final Kinetic Energy

For the fixed starting kinetic energy $T_0 = mU_0^2$ and the fixed final depth, the following formulae have been obtained, [37]

$$\bar{U}^* = e^{-1.5} \approx 0.223,\ S^* = \frac{3b\sigma^{1/3}}{eC_V},\ b = \frac{\rho_b^{2/3} T_0^{1/3}}{\rho\left[g(h_f + 10)\right]^{1/3}} \tag{14}$$

For all the four isoperimetric problems the optimal value of the optimal final cavitation number σ^* must be as small as possible and the optimal hull shape must coincide with the cavity shape at this cavitation number. For example, when $\sigma^* = 0.01$, then $S^* \approx 152b$. The optimal values of the body mass can be obtained for the cases with the fixed starting and final kinetic energy respectively:

$$m^* \approx \frac{0.00025\, T_0}{g(h_f + 10)},\ m^* \approx \frac{0.005\, T_f}{g(h_f + 10)}$$

Then the optimal body volume, length, caliber and cavitator radius can be calculated with the use of Eq. 5. To determine the maximum range in the problems with the fixed starting depth, it is necessary to solve the non-linear equations, similar to (11).

Thus, the presented simple analysis makes it possible to obtain the analytical relations for the optimal supercavitating model parameters and the optimal starting parameters. The obtained relations are in good agreement with the more accurate computer calculations.

5.1.5 Maximum Range for Slender Cavitators

It must be noted that the results for the maximum range and the optimal value of the final cavitation number presented in four previous subsections don't depend on the cavitator shape. The shape of the non-slender cavitator is important only for determining its optimal radius. According to the formulas (5) the cavitators with the greater values of C_x must have the smaller optimal radius and vice versa.

It is interesting to investigate the efficiency of slender cavitators. Some results are presented in [38]. The first approximation equation (6) and analytic formulas for the drag coefficient [17] were used to estimate the maximum range in the first isoperimetric problem. To precise these estimations the non-linear approach of the paper [30] was used. The results of calculation of the value $D^2\sigma/(4C_xR_n^2)$ for the cones with different half-angles θ are presented in Table 1. According to the Garabedian formulas (5), this ratio is equal to 1.0 for non-slender cavitators. The calculations show that slender cavitators can ensure the same or even grater values of the range. The same conclusion can be done for all isoperimetric problems 1–5 and for the problems with the fixed momentum and kinetic energy from Sects. 5.1.3 and 5.1.4. Therefore, the effectiveness of slender and non-slender cavitators is comparable.

5.1.6 The Maximum Range Estimations for the Hulls Without Separation and Cavitation

The comparison of the effectiveness of the supercavitating and unseparated hulls can be easily done for the isoperimetric problems 1–5. It is enough to compare the ranges of a supercavitating body and an unseparated one, when their velocities change from the starting value U_0 to the final speed of washing off U. Since the difference in these velocities is not very large (see, for example, (10)) and the volumetric drag coefficients are slightly dependent on the velocity (see (3)), the drag coefficients of the unseparated flow pattern at the velocity U have to be compared with the value presented by the formula (4). Therefore, for the isoperimetric problems No. 3, the supercavitating flow pattern is preferable, when the volumetric Reynolds number $\mathrm{Re}_V < 10^7$. This case corresponds to the very small hull volume (from 10 to 1 cm^3 for the velocities in the range 500–1,000 m/s).

Similar conclusions can be done for other isoperimetric problems as well. In [38] the critical values of the Reynolds number are calculated. In particular, for the first isoperimetric problem the critical value can be estimated as follows: $\mathrm{Re}_D^{(cr)} \approx 500000$. Thus, for the velocity range 100–1,000 m/s the supercavitation have to be used for the hulls of 5–0.5 mm diameter only. It must be noted, that the

Table 1 The value of $D^2\sigma/(4C_xR_n^2)$ for the cones with different half-angles θ

σ	0.1	0.05	0.025	0.01
$\theta = 15°$	1.115	1.094	1.072	1.031
$\theta = 10°$	1.116	1.094	1.078	1.064
$\theta = 5°$	1.128	1.107	1.089	1.066
$\theta = 3°$	1.113	1.108	1.097	1.079

aspect ratio of the optimal unseparated shape has to be minimal for the isoperimetric conditions 1 and 5. For other problems the optimal aspect ratio must be maximal (see [38]).

5.1.7 The Range Increase with the Use of a Cavitator with Changeable Shape and Diameter

The initial stages of the supercavitating inertial movement correspond to higher velocities and greater cavities in comparison with the final stages. Therefore, there is no need to use a large cavitator with a high drag for the initial stages of movement. The range can be increased with the use of cavitators with the changeable diameter or shape. The corresponding estimations have been done in [37]. For the horizontal movement with the final cavitation number $\sigma = 0.01$, the optimal value of the parameter $c = \alpha\sqrt{\omega} = 1.438$ was calculated. The values of the cavitator radius and its drag coefficient for the first stage are connected with the final characteristics as follows: $R_n = R_{n1}\alpha,\ \alpha \geq 1,\ C_x = C_{x1}\omega,\ \omega \geq 1$. The increase of the range is approximately 64%.

5.1.8 The Range Increase by Means of Propulsion

A propulsor evidently increases the range of a vehicle. If a thrust is constant and equal to the cavitator drag, the following formula can be used to estimate the additional value of range, [38]

$$\Delta S = \frac{2Q\Delta m}{\pi\rho U C_x R_n^2} \tag{15}$$

Where Δm is the mass of a fuel, Q is its specific momentum. According to the formula (15) the additional range is the inverse proportion to the velocity, therefore the propulsor is effective for the final stages of the vehicle movement with minimal velocities (before washing off). Formula (15) was analyzed in [38] for the supercavitating vehicles with different isoperimetric conditions. Since the range in this case of steady motion is independent from the body mass, but depends on the mass of the fuel, the most interesting problems can be divided into two groups:

1. The mass of the fuel is fixed
2. The density of the fuel ρ_f and its part in the total hull volume K_f are fixed

For both groups the standard additional limitations of the fixed caliber, length or volume must be added. Thus, the total number of the possible isoperimetric problems is 6. One of them (the density of the fuel ρ_f, its part in the total hull volume K_f and the hull length are fixed) has been considered in [40]. The solutions of other problems can be found in [38].

For example, when the mass of the fuel and the body caliber are fixed, the formulas (15) and (5) yield

$$\Delta S = \frac{8Q\Delta m}{\pi\rho U D^2\sigma}. \tag{16}$$

Therefore, the maximum range can be achieved with the use of minimal possible value of the cavitation number (for example, $\sigma = 0.01$). According to (16), the optimal range doesn't depend on the shape of the cavitator, since there is no dependence on its drag coefficient (similar to the case of inertial motion). The same conclusions can be done for all six isoperimetric problems (see [38]).

For the first group of isoperimetric conditions and the fixed values of the body caliber or its length, the optimal hull shape must be inscribed in the cavity corresponding to $\sigma = 0.01$; its caliber must coincide with the cavity caliber or its length must coincide with the cavity length accordingly. For other four isoperimetric conditions the optimal hull must coincide with the cavity shape at $\sigma = 0.01$.

It must be noted, that in the case of the unseparated hulls, their optimal aspect ratio can be minimal. Such situation occurs for the following two problems: (1) the mass of the fuel and the body caliber are fixed; (2) the density of the fuel ρ_f, its part in the total hull volume K_f and the body length are fixed. It is impossible to achieve an unseparated flow pattern with the use of very thick hulls (with a near to zero aspect ratio), but a special shape can be used to ensure the flow without separation and cavitation (see [25, 26]). Other four problems need the maximal value of the unseparated hulls aspect ratio.

5.2 Very Small Cavitation Numbers. Using the Nose of the Cavity Only

If the body mass, its caliber and aspect ratio are fixed, there is no need to investigate the case of the fixed body length. The volume of the hull located in the initial part of the cavity after a non-slender cavitator can be estimated as a cone volume (see Fig. 1c,d); therefore there is no need to use the isoperimetric condition with the fixed volume. Thus, it is necessary to investigate the first problem only from the list, presented in Sect. 5.1.

Both the natural and the ventilation cavitation will be taken into account with the given value of the cavity pressure p_c at the final moment of the hull washing off (when the vehicle stops). The cavitation number can be rewritten as follows:

$$\sigma = \frac{2gh_2}{U_0^2\bar{U}^2}, \qquad \bar{U} = \frac{U}{U_0}. \tag{17}$$

The final depth h and $h_2 = 10 + h - p_c$ are measured in meters.

5.2.1 Problems with the Fixed Final Depth

If in addition to the body mass, caliber, aspect ratio, final depth, its final velocity is also fixed, the final cavitation number will be also fixed (see (17)). Then the Garabedian formulae (5) allow calculating the hull shape and the cavitator diameter. Equation 9 shows that maximum range corresponds to the maximal starting

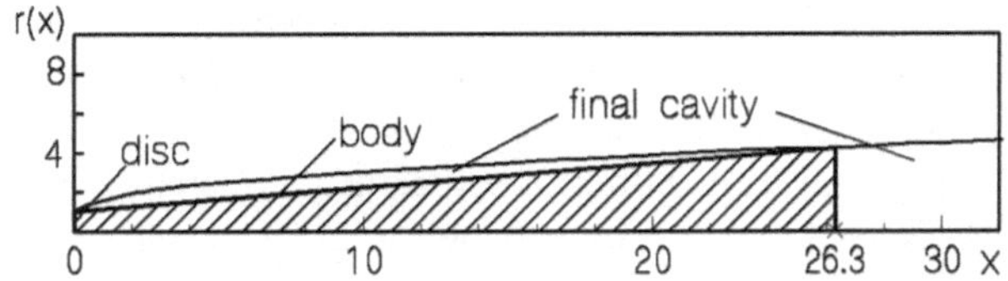

Fig. 14 An example of the optimal body shape for $\lambda_m = 3, \bar{H} = 0.001$

velocity. The same trivial solution will be in the case of complete cavity volume using (Fig. 1b) and for all five problems listed in Sect. 5.1.

If instead of mass, the initial body kinetic energy T_0 is fixed, then the optimal values of final velocity and body mass can be calculated (see details in [32])

$$\bar{U}^* = e^{-0.5} \approx 0.607, \qquad m^* = \frac{2T_0}{eU^2}.$$

The case of the fixed initial velocity is more difficult. But if the cavity volume is used completely (Fig. 1b), the same relationship (see Eq. 10) was obtained in [36]. For other four listed isoperimetric conditions, other relationships for the optimal velocities ratio were obtained in [36] with the use of the 1b pattern. The case of the partial using of the cavity volume (Fig. 1c,d) needs solving the non-linear equations and depends on the dimensionless parameter $\bar{H} = gh_2/U_0^2$.

The results for the dimensionless maximum range $\bar{S}^*$

$$\bar{S}^* = \frac{S^* g h_2 \rho D_b^2}{mU_0^2}$$

(solid lines) and optimal final cavitation number σ^*(dashed lines) are presented in Fig. 13 for different values of the hull aspect ratio. The range increases with the increasing of the aspect ratio, but the differences are sufficient for the very small values of the parameter $\bar{H}$ only. For $\bar{H}$>0.001 and λ_m>15 the obtained solution is practically independent of $\bar{H}$ and coincides with the results for the flow pattern 1b reported in [36].

The conclusion that the optimal hull caliber must coincide with the maximum final cavity diameter (see [36]) is no more valid for the case of very high velocities (small values of $\bar{H}$). To illustrate this fact, an example of optimal shape is shown in Fig. 14. The parameters of this supercavitating hull are $\lambda_m = 3$, $\bar{H} = 0.001$. The optimal range $\bar{S}^* \approx 0.083$ exceeds the ranges of any other hull with the same values of $\bar{H}$ and λ_m. For example, if the hull caliber coincides with the maximum final cavity diameter, then $\bar{S}^* \approx 0.063$only.

5.2.2 Problems with the Fixed Initial Depth

When the initial depth $h_0$is fixed, the relation for h_2 can be rewritten as follows: $h_2 = h_1 - S\sin\gamma$, $h_1 = 10 + h_0 - p_c$. It means that the cavitation number and the solutions of the problem will depend on the angle γ. The nonlinear equations make the search for the optimal solution more complicated.

If in addition to the body mass, caliber, aspect ratio, initial depth, its final velocity is also fixed, the maximum range corresponds to the maximal starting velocity. The same trivial solution will exist in the case of complete cavity volume using (Fig. 1b) and for all five problems listed in Sect. 5.1.

If instead of final velocity, the initial one is fixed, the solution is not trivial and depends on the dimensionless parameters $\bar{N}$, $\bar{h}_1$

$$\bar{N} = \frac{mg}{\rho U_0^2 D_b^2}, \qquad \bar{h}_1 = \frac{h_1 D_b}{U_0}\sqrt{\frac{\rho g}{m}}.$$

The results for the dimensionless maximum range $\bar{S}^*$

$$\bar{S}^* = \frac{S^* D_b}{U_0}\sqrt{\frac{\rho g}{m}}$$

(solid lines) and optimal final cavitation number σ^*(dashed lines) are presented in Figs. 15–17 for different values of the hull aspect ratio. The range increases with the increasing of the aspect ratio, but the differences are sufficient for the very small values of the parameter $\bar{N}$ only. For $\bar{N}>3 \cdot 10^{-7}$ and $\lambda_m>15$ the obtained solution is independent of $\bar{N}$ and coincide with the results for the flow pattern 1b obtained in [38].

The numerical analysis showed that for $\gamma>0$ the solution exist only for the values of $\bar{h}_1$which are greater than the critical one $\bar{h}_1^{(cr)}$. The situation is similar to the flow pattern 1b investigated in [38]. It means that for smaller values of the initial depth the body can reach the free water surface without a loss of the supercavitating flow pattern. The critical values $\bar{h}_1^{(cr)}$can be seen in Fig. 17. Increasing the hull aspect

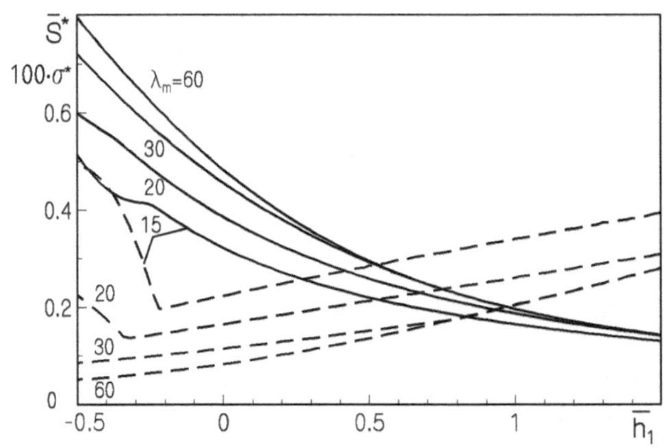

Fig. 15 Dependencies of the maximum range and optimal final cavitation number for different values of the hull aspect ratio at $\bar{N} = 10^{-7}$, $\gamma = -90°$

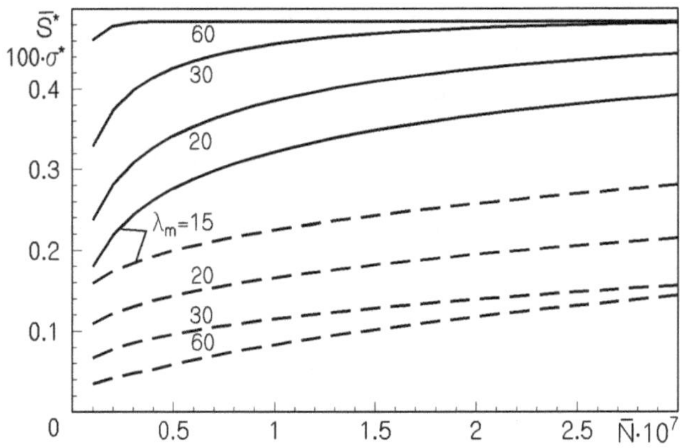

Fig. 16 Dependencies of the maximum range and optimal final cavitation number for different values of the hull aspect ratio at $h_1 = 0$, $\gamma = -90°$

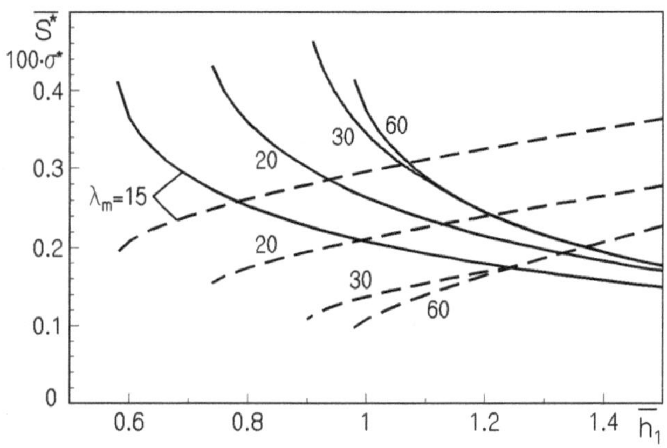

Fig. 17 Dependencies of the maximum range and optimal final cavitation number for different values of the hull aspect ratio at $\bar{N} = 10^{-7}$, $\gamma = 90°$

ratio increases $\bar{h}_1^{(cr)}$, which tends to the value (13) obtained in [38] for the flow pattern 1b.

6 Non-standard Hulls and Cavities

If a part of a vehicle is covered by the cavity, which closes itself (without any fictitious closing rigid body or re-entrant jet, see Fig. 2), the hull's pressure drag has to be near to zero (due to d'Alembert paradox). Only the friction in the boundary-layer determines the body drag. The skin-friction drag on such vehicle can be reduced, since the large part of its surface has no contact with the water (an advantage of the supercavitating flow pattern). On the other hand, such vehicle has no high pressure drag (a typical disadvantage of the supercavitation flow pattern).

Due to the smaller area of contact with the water, the skin-friction drag coefficients cab be estimated as follows for the laminar and the turbulent boundary-layer respectively, [41]:

$$C_V = \frac{4.708}{\sqrt{\mathrm{Re}_V}}\sqrt{\frac{V_b}{V}}, \tag{18}$$

$$C_V = \frac{0.073\ {V_b}^{6/7}}{{\mathrm{Re}_V}^{1/7} V^{13/21}}, \tag{19}$$

where V_b is the volume of the body's part wetted by water.

In order to realize the flow pattern shown in Fig. 2, a special investigation has been done in [28]. It was shown that shapes of the axisymmetric slender cavities can be not only elliptical. The cavity longitudinal cross section can be also a parabola, a concave and convex hyperbola and even a straight line (at one specific value of the cavitation number). The last case shows, that the slender body theory is applicable up to the end of the straight line cavity (in comparison with the elliptical cavities with the infinite slope at the blunt trailing edge). Therefore, such cavity can close itself and does not need any artificial closing scheme.

To support this fact the non-linear calculations have been performed in [28] with the use of sources and doublets located on the axis of symmetry. Their intensity was chosen to satisfy the constant pressure condition on the cavity surface. An example is presented in Fig. 18 with $V_b = 2.6 \cdot 10^{-4}$, $V = 5.5 \cdot 10^{-4}$, dimensionless volumes are based on the cubic body length. Equation (18) gives the drag diminishing of 31% (in comparison with the unseparated flow pattern $V_b = V$). For the pure turbulent boundary-layer (Eq. 19) the advantage is 47%. Formula (19) yields the estimation $C_V \approx 5 \cdot 10^{-4}$ for the body shown in Fig. 18. This value is 14 times less than the volumetric drag of the underwater apparatus "Dolphin"

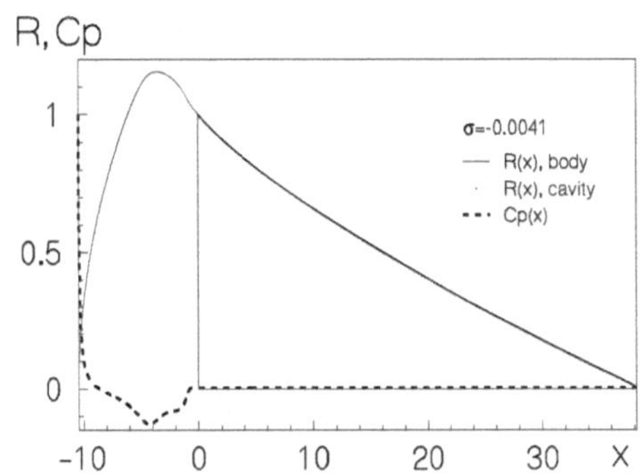

Fig. 18 Axisymmetric cavitator ($x < 0$) and cavity, which closes itself ($x > 0$). Shape and pressure distribution

measured at $\mathrm{Re}_V = 8.5 \cdot 10^6$(see [42]). These small values of C_V let the untypical supercavitating hull shown in Fig. 18 be effective even in the cases of horizontal motion (when the buoyancy force must be compensated). From Fig. 8 it can be seen that the critical volume of the effective supercavitating hull can be rather large for $C_V \approx 5 \cdot 10^{-4}$ and high enough values of the aerodynamic effectiveness $k = C_y/C_x$.

The diminishing V_b/V leads to the drag reduction. Nevertheless, the short cavitators have more deep pressure minimum on their surface. This fact can cause separation (and cavitation) upstream to the point $x = 0$ and another flow pattern with a large pressure drag. The separation behavior is very important for such flow pattern and has to be investigated in a water tunnel. Some results of wind tunnel tests with unseparated shapes are presented in [25, 26].

Conclusions

The value of the volumetric drag coefficient and the similar coefficients, based on the squared values of the length and the caliber, can sufficiently be reduced at cavitation number less than 0.01. The smallest values of these drag coefficients correspond to the largest aspect ratios and the slender cavitators. Comparison of the supercavitating and unseparated flow patterns showed the existence of the critical values of the volume and dimensions. The supercavitating flow pattern is preferable for the values of these parameters smaller than critical ones. For the horizontal supercavitation motion, the necessity of the buoyancy force compensation sufficiently diminishes the critical values of the vehicle volume or its dimensions, which achieve maximum at a certain value of the motion velocity. In the case of the base cavity, the comparison the supercavitating and the unseparated flow patterns is presented. The critical value of the body volume has a maximum at a certain value of the movement velocity and drastically increases with the aspect ratio increasing.

Maximum range problems are considered for the supercavitating motion of the axisymmetric body on inertia under an arbitrary angle to horizon. Different isoperimetric problems were formulated and solved with the fixed values of the body mass, kinetic energy, aspect ratio and caliber. Two dimensionless parameters are proposed which influence the solution. At small values of these parameters the optimal body shapes may use the nose part of the cavity only. Analytic and numeric solutions for the maximal range and the optimal body shapes are obtained. It was shown that infinite small exceeding of some critical value of the initial depth can cause a jump of the range (body comes to the water surface without loosing the supercavitating flow pattern).

The presented analysis showed that supercavitating hulls can be successfully used to diminish the drag especially at very small cavitation numbers. Supercavitation may be preferable for not very large hulls which use whole cavity space or its nose part only. Non-standard cavitators and cavities shown in Fig. 2 can be successfully used to reduce the total drag.

References

1. Logvinovich GV. Hydrodynamics of Flows with Free Boundaries. Kiev: Naukova Dumka; 1969. (In Russian). English translation: Halsted; 1973.
2. Buyvol VN. Slender cavities in flows with perturbations. Kiev: Naukova Dumka; 1980 (In Russian).
3. Logvinovich GV, Buyvol VN, Dudko AS, et al. Free boundary flows. Kiev: Naukova Dumka; 1985 (In Russian).
4. Savchenko YuN. Supercavitating object propulsion. RTO-AVT/VKI Special Course on Supercavitating Flows; February 12–16, 2001. VKI, Brussels; 2001. (Belgium).
5. Savchenko YuN. On motion in water in supercavitation flow regimes. Gidromehanika. 1996;70:105–15 (In Russian).
6. Savchenko YuN, Vlasenko YuD, Semenenko VN. Experimental study of high-speed cavitated flows. Int J Fluid Mech Res. 1999;26(3):365–74.
7. Savchenko YuN, Semenenko VN, Putilin SI. Unsteady supercavitated motion of bodies. Int J Fluid Mech Res. 2000;27(1):109–37.
8. Savchenko YuN. Perspectives of the supercavitation flow applications. Proceedings of the International Conference on Superfast Marine Vehicles Moving Above, Under and in Water Surface (SuperFAST'2008), 2–4 July 2008. St. Petersburg; 2008. ISBN 5-88303-393-8.
9. Savchenko YuN, Semenenko VN, Putilin SI, et al. Designing the high-speed supercavitating vehicles. International Conference on Fast Sea Transportation (FAST'2005), June 2005. St. Peterburg; 2005.
10. Semenenko VN. Some problems of supercavitating vehicle designing. Proceedings of the International Conference on Superfast Marine Vehicles Moving Above, Under and in Water Surface (SuperFAST'2008); 2–4 July 2008. St. Petersburg; 2008. ISBN 5-88303-393-8.
11. Logvinovich GV, Serebryakov VV. On the methods of calculating a shape of the slender axisymmetric cavities. Gidromehanika. 1975;32:47–54 (In Russian).
12. Vlasenko YuD. Experimental investigations of supercavitating regime of flow around self-propelled models. Int J Fluid Mech Res. 2001;28(5):717–33.
13. Savchenko YuN, Semenenko VN. Special features of supercavitating flow around polygonal contours. Int J Fluid Mech Res. 2001;28(5):660–72.
14. Putilin SI. Some features of a supercavitating model dynamics. Int J Fluid Mech Res. 2001; 28(5):631–43.
15. Nesteruk I. Investigation of slender axisymmetric cavity form in fluid with gravity. Izv AN SSSR MFG. 1979;6:133–6 (In Russian).
16. Nesteruk I. Form of slender axisymmetric cavity. Izv AN SSSR MFG. 1980;4:38–47 (In Russian).
17. Nesteruk I. Some problems of axisymmetric cavity flows. Izv AN SSSR MFG. 1982;1:28–34 (In Russian).
18. Nesteruk I. The slender axisymmetric cavity form calculations based on the integral-differential equation. Izv AN SSSR MFG. 1985;5:83–90 (In Russian).
19. Goman OG, Semenov YA. Oblique entry of a wedge into an ideal incompressible fluid. Fluid Dynam. 2007;42(4):581–90.
20. Faltinsen OM, Semenov YA. The effect of gravity and cavitation on a hydrofoil near the free surface. J Fluid Mech. 2008;597:371–94.
21. Savchenko YuN, Semenov YA. Hydrodynamic drag of a surface with the mixed boundary conditions. Prykladna Gidromehanika. 2005;7(2):54–62 (In Russian).
22. Savchenko YuN, Savchenko GYu. Efficiency estimation of the supercavitation using on the axisymmetric hulls. Prykladna Gidromehanika. 2004;6(4):78–83 (In Russian).
23. Savchenko YuN, Savchenko GYu. Cylinder planning on the supercavity surface. Prykladna Gidromehanika. 2007;9(1):81–5 (In Russian).
24. Savchenko YuN, Savchenko GYu. Near-wall cavitation on a vertical wall. Prykladna Gidromehanika. 2006;8(4):53–9 (In Russian).

25. Nesteruk I. The problems of drag reduction in high speed hydrodynamics. The International Summer Scientific School "High Speed Hydrodynamics"; June 16–23, 2002. Cheboksary; 2002. p. 351–9.
26. Nesteruk I. Drag reduction in high-speed hydrodynamics: supercavitation or unseparated shapes. CAV2006; 2006. Netherlands.
27. Nesteruk I. Drag calculation of slender cones using of the second approximation for created by them cavities. Prykladna Gidromekhanika (Appl Hydromech), Kyiv. 2003;5(1):42–6 (In Ukrainian).
28. Nesteruk I. Partial cavitation on long bodies. Prykladna Gidromekhanika (Appl Hydromech), Kyiv. 2004;6(3):64–75 (In Ukrainian).
29. Garabedian PR. Calculation of axially symmetric cavities and jets. Pac J Math. 1956; 6(4):611–84.
30. Nesteruk I. Simulation of axisymmetric and plane free surfaces by means of sources and doublets. Prykladna Gidromekhanika (Appl Hydromech), Kyiv. 2003;5(2):37–44 (In Ukrainian).
31. Nesteruk I. Drag diminishing of long axisymmetric high-speed bodies. Prykladna Gidromekhanika (Appl Hydromech), Kyiv. 2009;11(2):55–67 (In Ukrainian).
32. Manova ZI, Nesteruk I, Shepetyuk BD. Optimization problems for high-speed supercavitation motion on inertia with the non-slender cavitators. Prykladna Gidromekhanika (Appl Hydromech), Kyiv. 2009;11(4):54–9 (In Ukrainian).
33. Gieseke TJ. Toward an optimal weapon system utilizing supercavitating projectiles. International Conference on Cavitation "Cav2001", Pasadena; 2001, Session B3.002.
34. Serebryakov VV. The models of the supercavitation prediction for high speed motion in water. International Summer Scientific School "High Speed Hydrodynamics". Cheboksary; 2002:71–92.
35. Serebryakov VV, Kirshner IN, Scherr GH. Some problems of high speed motion in water with supercavitation for sub-, trans- and supersonic mach numbers. Proceedings of the X International scientific school "High-speed hydrodynamics" and International conference «Hydromechanics. Mechanics. Power-plants» (to the 145-th anniversary of academician A.N.Krylov). Moscow/Cheboksary: Cheboksary department of Moscow State Open University; 2008. p. 73–104. ISBN 978-5-902891-35-2.
36. Nesteruk I, Semenenko VN. Problems of optimization of range of the supercavitation inertial motion at the fixed final depth. Prykladna Gidromekhanika (Appl Hydromech), Kyiv. 2006; 8(4):33–42 (In Ukrainian).
37. Nesteruk I, Savchenko YuN, Semenenko VN. Range optimization for supercavitating motion on inertia. Rep Ukrainian Acad Sci. 2006;8:57–66 (In Ukrainian).
38. Nesteruk I. Range maximization for supercavitation inertial motion with the fixed initial depth. Prykladna Gidromekhanika (Appl Hydromech), Kyiv. 2008;10(3):51–64 (In Ukrainian).
39. Nesteruk I. Hull optimization for high-speed vehicles: supercavitating and unseparated shapes. International Conference SuperFAST2008, July 2–4, 2008. St. Petersburg; 2008.
40. Savchenko YuN. Investigations of supercavitation flows. Prykladna Gidromekhanika (Appl Hydromech), Kyiv. 2007;9(2–3):150–8 (In Russian).
41. Buraga OA, Nesteruk I, Savchenko YuM. Comparison of the slender axisymmetric bodies drag by unseparated and supercavitation flow patterns. Prykladna Gidromekhanika (Appl Hydromech), Kyiv. 2002;4(2):3–8 (In Ukrainian).
42. Lorant M. Investigation into High-Speed of Underwater Craft. Naut Mag. 1968;200(5):273–6.

Hydrodynamic Characteristics of a Disc with Central Duct in a Supercavitation Flow

G. Yu. Savchenko

Abstract
Supercavitation flow past a disc cavitator with a round hole at its center is studied experimentally. The cavity drag of a family of discs with different values of the inner-to-outer diameter ratio $\frac{d}{D}$ is measured in a water tunnel. The results are presented as the drag coefficient and the factor of added mass versus the diameter ratio and compared with the solution of the plane problem for two symmetrical plates and with experimental results by other authors.

1 Introduction

A ducting cavitator is of considerable practical interest as a water disc-duct inlet or a braking device. Whereas for flow past simple configurations such as a cone, a disc, and a sphere the results of experimental studies and theoretical calculations are well known [1, 2], few results have been reported for the disc with a hole.

In 1959, Tseitlin (Central Aerohydrodynamic Institute (TsAGI)), proposed that the solution of the plane problem of flow past two flat plates be extended to flow past the disc with a hole [3]. In 1994, Deinekin (Institute of Hydromechanics of the National Academy of Sciences of Ukraine) numerically solved the problem of axisymmetric flow past a ducting cavitator [4]. However, the verification of these solutions and comparison of the theoretical results with experimental data are still topical, which is the aim of this work.

G.Yu. Savchenko (✉)
Institute of Hydromechanics of the National Academy of Sciences of Ukraine, Kyiv, Ukraine
e-mail: Georgiy_Savchenko@yahoo.com

I. Nesteruk (ed.), *Supercavitation*,

DOI 10.1007/978-3-642-23656-3_6, © Springer-Verlag Berlin Heidelberg 2012

2 Experimental Procedure

Experiments were conducted at the Hydrodynamic Laboratory, the Institute of Hydromechanics of the National Academy of Sciences of Ukraine, on a water tunnel with an open working area measuring 0.34 × 0.34 × 2.0 m at a flow velocity of 8.9 m/s. Duralumin discs of outer diameter $D = 47$ mm with different diameters d of their center hole (Fig. 1) were used in experiments. The discs were fastened using a ∅ 1.2 mm flexible cable with four guy ropes as shown in Figs. 2 and 3.

Model	D, mm	d,mm	$\frac{d}{D}$	h,mm
M47/1	47	37	0.79	5.0
M47/2	47	32	0.68	7.5
M47/3	47	27	0.57	10.0
M47/4	47	21	0.45	13.0
M47/5	47	9	0.19	19.0
M47/6	47	0	0	23.5

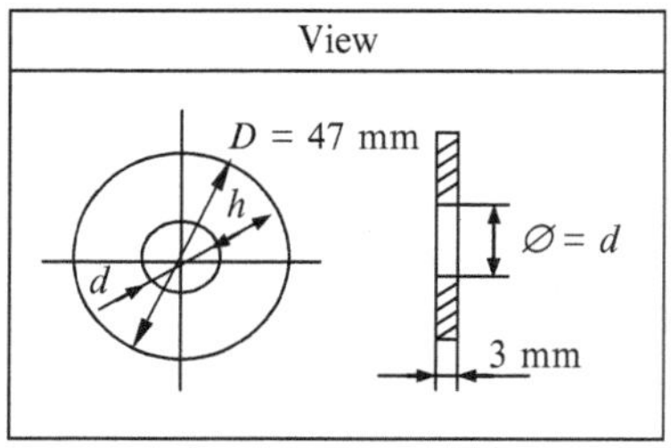

Fig. 1 The disc cavitator with different diameters of the center hole

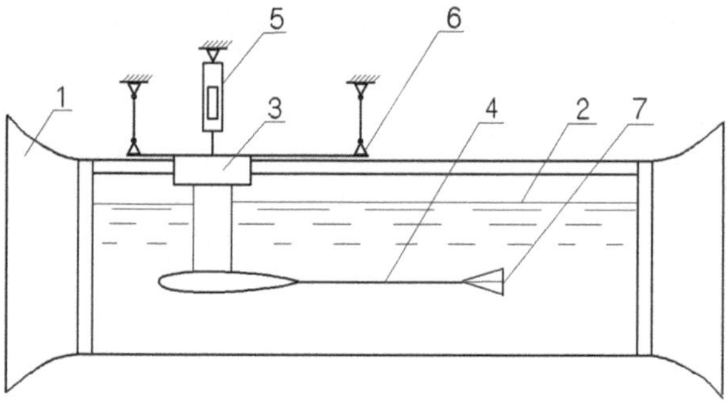

Fig. 2 1 – water tunnel working area; 2 – free water surface; 3 – model fastening fairing; 4 – model fastening flexible cable; 5 – dynamometer; 6 – hinged suspension of fairing; 7 – the model – disc cavitator

Figure 2 shows a schematic of the water tunnel working area, 1, in which a measuring bench is mounted. The bench includes a fairing 3, through which a flexible cable 4 is threaded to fasten a disc cavitator 7. The other end of the cable is fastened to a dynamometer 5.

During the tests, the fairing 3 with the model 7 was sunk into the water flow using a hinged suspension 6. While traversing the free boundary, the model 7 entrained some air, and for some time it was in the flow in the regime of developed cavitation (see Fig. 3b). 10–20 s later, when the air from the cavity had been entrained by the flow, the regime of continuous flow set in (see Fig. 3a). The hydrodynamic resistance of the model in the regimes of supercavity and continuous flow was measured using a type DFC 60–11 electronic digital dynamometer and a storage oscilloscope.

During the experiments, the flow velocity in the water tunnel was kept constant $V_\infty = 8.9\ \mathrm{m/s}$, which was provided by keeping a constant water level of 4.04 m in the head tank.

3 Experimental Results

The test results for the series of models are plotted in Fig. 4 as the drag coefficient C_x versus a dimensionless parameter – the ratio of the inner diameter d of the cavitator to its outer diameter D. The drag coefficient is calculated as

$$C_x = \frac{2X}{\rho V_\infty^2 S}, \tag{1}$$

where ρ is the water density (1,000 kg/m^3 at 15°C); $V_\infty = 8.9\ \mathrm{m/s}$ is the inflow velocity; $S = \frac{\pi}{4}\left(D^2 - d^2\right)$ is the cavitator surface area according to the dimensions in Fig. 1; and X [N] is the hydrodynamic resistance of the cavitator model.

3.1 Continuous Flow

The plots show that the model drag coefficient in the regime of continuous flow tends to $C_x \approx 2.0$ as the inner diameter increases to $d/D \to 1$, which corresponds to the drag coefficient of a flat plate [2].

Thus it may be thought that at d/D close to one the model drag coefficient is close to two

$$C_x = 2.0; \quad \frac{d}{D} > 0.85. \tag{2}$$

At $d/D \to 0$ in the regime of continuous flow the drag coefficient is close to that of a disc

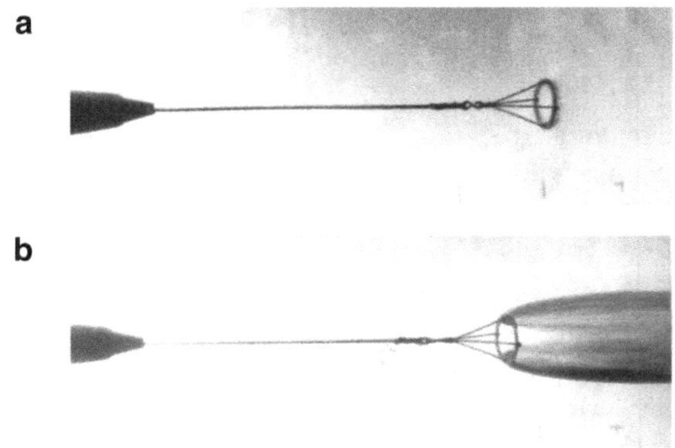

Fig. 3 Continuous (**a**) and cavitation (**b**) flow past a model

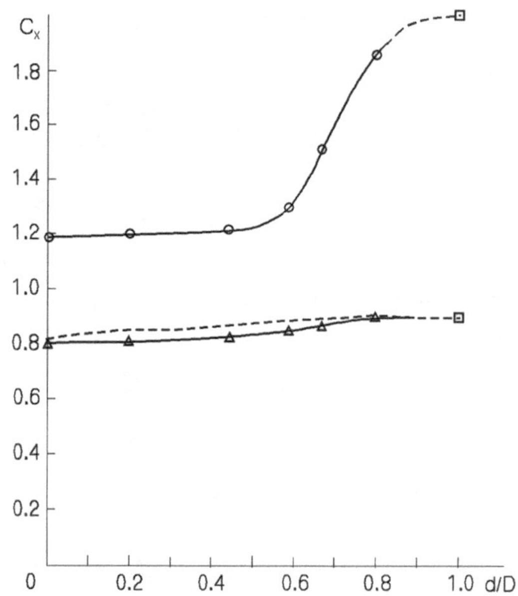

Fig. 4 ○ – continuous flow; Δ – cavity flow; - - - - calculation [3]

$$C_x = 1.2; \quad \frac{d}{D} < 0.45. \tag{3}$$

3.2 Supercavitation Flow

In the regime of supercavitation flow where a supercavity forms downstream of the model (Fig. 3b), the drag coefficient varies through the range $0.79 < C_x < 0.89$ if $0 < d/D < 1$ (see Fig. 4). The dashed line in Fig. 4 shows the results of theoretical calculation of the drag coefficient of a disc with a center hole based on the solution of the plane problem of flow past two symmetrical plates [3].

Since the theoretical dependence is too complex and cumbersome, we propose the interpolation formula based on the experimental results (Fig. 4), which is more convenient for practical use

$$C_x\left(\sigma;\frac{d}{D}\right) = \left(0.8 + 0.08\frac{d}{D}\right)(1+\sigma). \tag{4}$$

Equation 4 may also be represented as the result of linear interpolation between the drag coefficient C_{xd} of a disc and the drag coefficient C_{xp} of a flat plate at $\sigma = 0$

$$\begin{aligned} C_x\left(\sigma;\frac{d}{D}\right) &= \left[C_{xd} + (C_{xp} - C_{xd})\frac{d}{D}\right](1+\sigma) \\ &= \left(0.82 + 0.06\frac{d}{D}\right)(1+\sigma), \end{aligned} \tag{5}$$

where $C_{xd} = 0.82(1+\sigma)$ is the drag coefficient of a disc [3]; $C_{xp} = \frac{2\pi}{4+\pi}(1+\sigma)$ is the drag coefficient of a flat plate [5]; $\sigma = \frac{2(P_0 - P_C)}{\rho V_\infty^2}$ is the cavitation number; and P_c, P_0 are the pressure in the cavity and in the flow, respectively.

A comparison shows that the disc drag coefficient obtained in the experiment is 10% smaller than the theoretical calculated value for C_{x0} at small values of σ. However, a comparison with results obtained in water tunnels shows that this decrease in C_{x0} is typical for water tunnels [1].

4 Determination of the Added Mass of a Disc with a Hole

4.1 Continuous Flow

The hydrodynamic force acting on a disc executing an accelerated motion has a component caused by its added mass [2, 6]. As known, the added mass of a disc in a continuous flow is (e.g., [2, 6])

$$m^* = \frac{8}{3}\rho R^3, \tag{6}$$

and the added mass of a flat plate is (e.g., [7])

$$m^* = \pi\rho A^2 L, \tag{7}$$

where $2A$ and L are the plate width and length, respectively, the length being greater than the width:

$$2A = h; \quad L > 2A; \quad 2R = D.$$

If a cavitator with an hole is thought of as a plate curved into a ring with dimensions $A = \frac{D-d}{2}$; $L = \pi \frac{D+d}{2}$, then, according to (7), the added mass of this ring will be

$$m^* = \frac{\pi^2}{4} \rho R^3 \left(1 - \frac{d}{D}\right)^2 \left(1 + \frac{d}{D}\right) \tag{8}$$

The coefficient of added mass may be written as

$$K_m = \frac{m^*}{\rho R^3} = \frac{\pi^2}{4} \left(1 - \frac{d}{D}\right)^2 \left(1 + \frac{d}{D}\right) \tag{9}$$

At $d = 0$, Eq. (9) must give the value of K_m for a disc of radius $R = D/2$.

When $K_m = \frac{\pi^2}{4} = 2.5$ obtained from (9) for a disc $(d = 0)$ is compared with the true value of K_m given by Eq. (6)

$$K_m = \frac{8}{3} = 2.67, \tag{10}$$

we can see that they differ by 7% only.

Figure 5 shows the factor of added mass $K_m(d/D)$of a disc with a hole calculated by Eq. 9 for the family of models under study.

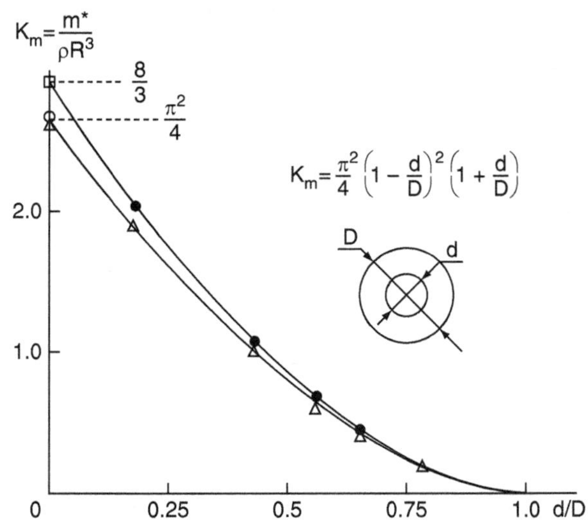

Fig. 5 ●– continuous flow
Δ – cavitation flow

4.2 Supercavitation Flow

In assessing the added mass in a supercavitation flow, it is important to keep in mind that the disc contacts the liquid on one side only. Theoretical estimates for a disc and a plate floating on a horizontal water surface are [8]:

$$\text{disc}: \ m^* = \frac{4}{3}\rho R^3;$$

flat plate of width $2A : m^* = \frac{\pi}{2}\rho A^2$.

To assess the added mass of a disc in a supercavitation flow, special experiments were conducted, and they gave a value close to that for the regime of continuous flow [6, 7, 9] $K_m = 2.5$, which differs from 10 by 6.4%.

Since the coefficient $\frac{\pi^2}{4}$ in Eq. (9) gives a close value for a disc ($d = 0$),

Equation 9 can be used in determining the factor of added mass of a flow passage disc in a supercavitation flow (curve Δ in Fig. 5).

Conclusions

Simple formulas are proposed to calculate the drag coefficient $C_x\left(\sigma; \frac{d}{D}\right)$ (4) and factor of added mass $K_m\left(\frac{d}{D}\right)$ (9) of a ducting disc in a supercavitation flow. These formulas are convenient for use in computer programs and in engineering calculations of cavitating water intakes and braking devices.

References

1. Knapp R, Daily J, Hammitt F. Cavitation (in Russian). Moscow: Mir Publishers; 1974.
2. Devnin SI. Aerohydromechanics of bluff structures (in Russian). Leningrad: Sudostroenie; 1983.
3. Tseitlin MYu. On the pressure on two parallel plates in a jet flow (in Russian). TsAGI Transactions on Hydrodynamics. Moscow; 1959. p. 296–308.
4. Deinekin YuP. Cavity flow past flow passage bodies (in Russian). Gidromekhanika. 1994;68:74–8.
5. Shashin VM. Hydromechanics (in Russian). Moscow: Vysshaya Shkola; 1998.
6. Logvinovich GV. Initial motion of a body in a liquid with developed cavitation (in Russian). TsAGI Transactions. 1959. p. 3–39.
7. Zhuravlev YuV. Entry of a disc into a liquid at an angle to the free surface (in Russian). TsAGI Transactions; 1959. p. 227–32.
8. Gurevich MI. Impact of a plate in a separated-jet flow (in Russian). Moscow: Prikladnaya Matematika i Mekhanika. 1952;XIV(1):116–8.
9. Savchenko YuN. Hydrodynamic forces acting on a disc executing sinusoidal oscillations (in Russian). Izvestiya AN SSSR, Mekhanika Zhidkosti i Gaza. 1971;2:186–7.

Gas Flows in Ventilated Supercavities

Yu. N. Savchenko and G. Yu. Savchenko

Abstract
The paper is concerned with a gas flow inside a ventilated supercavity. It is found out that there exists a ring vortex flow inside a supercavity, which forms a re-entrant counterflow and free boundary disturbances. Attention is drawn to the specific character in which a gas flow inside a supercavity disturbs its free boundaries. Formulas are given to estimate the gas flow rate and the supercavity development time and to calculate the supercavity – moving object gap.

1 Introduction

The study of flows in ventilated supercavitation attracts considerable interest since forming ventilated supercavities on vessel hulls and bodies of revolution offers a several-fold reduction in drag thus enabling the moving object to travel in water at a far greater speed [1–3]. Gas injection can be used to control supercavity flows [4, 5].

In the literature, two types of gas entrainment from free ventilated supercavities have been pointed out: gas entrainment by vortex filaments and gas entrainment by periodically detaching portions (portion entrainment). Cox and Claiden [6] and Epshtein [3] attempted to develop a theory of gas entrainment by vortex filaments. Epshtein [3], Logvinovich [2], Krylov [7], and others studied gas entrainment experimentally.

Then attempts were made to relate the viscous resistance of gas flow inside a ventilated cavity to gas entrainment therefrom. Here, the works by Epstein [1, 8] and Spurk [9, 10] should be mentioned, where it was assumed that only the gas entrapped by the boundary layer on the free cavity boundary is entrained from the

G.Yu. Savchenko (✉)
Institute of Hydromechanics of the National Academy of Sciences of Ukraine, Kyiv, Ukraine
e-mail: Georgiy_Savchenko@yahoo.com

I. Nesteruk (ed.), *Supercavitation*,

DOI 10.1007/978-3-642-23656-3_7,

cavity. In the adopted theoretical scheme of gas entrainment, the boundary layer on the cavity boundary was assumed to be similar to a layer formed on a rigid wall.

In reality, the water and the gas flows in a ventilated cavity are interrelated. The gas flow via its relative velocity

$$V_e = V_c - V_g, \tag{1}$$

where V_c is the velocity on the supercavity boundary; V_g is the gas flow velocity, may affect the breaking of the free liquid cavity boundary, which in its turn affects gas entrainment from the cavity.

This work is aimed at elucidating the gas flow pattern inside ventilated supercavities and its effect on cavity gas entrainment.

2 Free Supercavity Boundary Disturbance

Experiments show that the gas entrainment rate from a free supercavity is governed by free boundary disturbances and closure conditions. Free boundary deformations may result under the action of gravity forces, when the Froude number Fr is small [1] and the trailing part of the supercavity starts to buoy, its cross-section deforming and transforming into two hollow vortices [2, 3].

Periodic variations of the instantaneous cavitation number may give rise to waves on the free supercavity boundary, thus resulting in the portion type of cavity gas entrainment [11, 12]. Small-scale supercavity boundary disturbances may be caused by turbulence in the inflow and cavitator vibrations.

Small-scale disturbances on the free supercavity boundary in the presence of a gas flow at the liquid–gas interface are due to the fundamental instability of a liquid surface described by Taylor and Helmholtz [13]. This instability manifests itself in waves on a free water surface and droplet detachment when the relative speed of the gas flow exceeds some critical value (12–13 m/s for water).

If the relative gas speed is far greater than the critical speed, the free boundary breaks to form a splash layer, wherein the droplets are accelerated to the gas flow velocity. In the cavity closure region, this splash flow forms a two-phase gas–water mixture, which is entrained continuously or intermittently into the wake in the form of individual portions and jets.

The literature on the effect of a wind flow on a water surface is quite voluminous ([13–16], etc.). As a result of the studies conducted, the water surface state has been related to the wind speed and action time [14]. However, supercavities feature gas flows with high speeds of 10^2–10^3 m/s and a characteristic action time of the order of 10^{-1} s while atmospheric processes exhibit speeds up to 25 m/s and a far longer action time of the order of 10^4–10^5 s. Nevertheless, the obtained results are of interest in that they allow one to find the minimum (critical) gas speed that corresponds to the onset of splashing.

Splashing as a kind of surface wave instability in the presence of a gas flow in the vicinity of interfaces was described by Taylor [17]. In this work, a definition of the instability wavelength λ_B is introduced: this wavelength is defined as the diameter $2r_k$ of detached droplets

$$\lambda_B = \frac{2 \cdot \pi \cdot \varsigma}{\rho_g \cdot V_g^2} \cdot f\left(\frac{\rho \cdot \varsigma^2}{\rho_g \cdot \eta^2 \cdot V_g^2}\right) = 2r_k = \frac{3\pi \cdot \gamma}{\rho_g V_g^2}, \tag{2}$$

where ζ is the liquid surface tension; V_g is the relative speed near the interface; ρ, ρ_g are the liquid and the gas density, respectively; and η is the liquid dynamic viscosity; for the water–air interaction

$$f\left(\frac{\rho \cdot \varsigma^2}{\rho_g \cdot \eta^2 \cdot V_g^2}\right) \approx 1.5.$$

Calculations by Eq. 2 show that droplets of diameter $d_\kappa = 2r_\kappa \approx$ 2–5 mm are formed at speeds higher than 10 m/s, and thus splashing by Taylor's mechanism at the water–air interface may be considered to start when

$$V_g \geq 10\ \mathrm{m/s}. \tag{3}$$

Another splashing mechanism involves the detachment of a capillary wave on a liquid surface (Helmholtz instability) [18]. Calculations show that this type of instability gives a close value of the critical speed $V_{g\ cr} = 11$ m/s. From other experimental data, which are reported in [19], the critical Weber number $\mathrm{We}_{\mathrm{cr}}$ corresponding to the onset of splashing is proposed

$$We_{cr} = \frac{\rho_g V_{g\ cr}^2}{\sqrt{\rho \cdot g \cdot \varsigma}} = 8.8. \tag{4}$$

At the following parameter values: air density ($T = 15^{\mathrm{o}}$) $\rho_g = 1.29$ kg/m^3, water density ($T = 15^{\mathrm{o}}$) $\rho = 1{,}000$ kg/m^3, and water surface tension $\varsigma = 0.074$ N/m, the critical gas speed will be $V_{g\ cr} = 13.5$ m/s. This value is close to theoretical estimates by the Helmholtz and the Taylor method, where $V_{g\ cr} = 11$ m/s.

3 Gas Flow Regimes in a Ventilated Supercavity

A ventilated supercavity is formed by gas injection into the supercavity at volume rate Q_{in}. In doing so, the gas balance equation for isothermal gas expansion inside the supercavity must be satisfied [11]

$$\frac{dW}{dt} = Q_{in} - Q_{out}, \tag{5}$$

where W is the supercavity volume, and Q_{out} is the gas outflow rate from the cavity.

In ventilated supercavitation, two characteristic regimes exist:

1. Formation of a ventilated supercavity, when the injection Q_{in} makes up for the increase of the supercavity volume $\frac{dW}{dt}>0$ and the gas is not entrained yet ($Q_{out} = 0$) because the supercavity is still unclosed. In this case, the gas balance is governed by the equation

$$\frac{dW}{dt} = Q_{in}, \tag{6}$$

while the cavity pressure P_c, the inflow velocity V_{∞}, and the cavitation number remain constant.

2. Steady flow regime, when the cavitation number $\sigma = \sigma_0$ is constant, the supercavity has reached its maximum size, and its volume remains constant $\left(\frac{dW}{dt} = 0\right)$. In this case, the gas entrainment rate is equal to the gas injection rate, and the gas balance equation has the form

$$Q_{in} = Q_{out}. \tag{7}$$

4 Supercavity Formation Regime

A similar regime takes place at the initial stage of penetration of cavitating objects through a free water surface or in passing through a solid underwater obstacle [4, 20]. In this regime, the velocity and pressure in a supercavity are constant, and the gas injection rate required for supercavity formation reaches its maximum value. The importance of this regime is due to the fact it minimizes the supercavity development time and allows one to assess the gas injection rate required for this purpose.

The injection process in the supercavity formation regime can be explained using the schematic shown in Fig. 1 where the cavitator starts moving from a

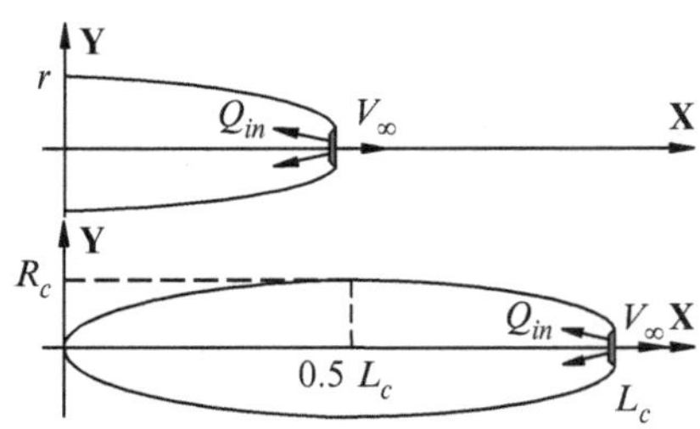

Fig. 1 Supercavity formation scheme with gas injection

solid impermeable wall at a constant velocity. The schematic shows the development of a supercavity to its full length L_c when the gas is injected into the cavity in the vicinity of the cavitator.

If the cavity contour is assumed to be near-elliptical, then its radius can be calculated as

$$R_x{}^2 = \frac{x \cdot (L_c - x)}{\lambda^2}, \tag{8}$$

$$\lambda = \sqrt{\frac{\ln \frac{1}{\sigma}}{\sigma}} = \frac{L_c}{2 \cdot R_c}$$

where λ is the cavity aspect ratio, and the cavity volume can be estimated as

$$W_c = \frac{2}{3}\pi \cdot L_c \cdot R_c^2. \tag{9}$$

In view of (8), the supercavity cross-section area is

$$S_c = \pi \cdot R_x^2 = \frac{\pi \cdot x \cdot (L_c - x)}{\lambda^2},$$

$$L_c = \frac{2 \cdot R_n}{\sigma}\sqrt{C_x \ln \frac{1}{\sigma}}; \quad R_c = R_n \sqrt{\frac{C_x}{\sigma}}$$

where L_c; R_c are the cavity length and midsection radius, R_{n}, C_{x} are the cavitator radius and drag coefficient, and the rate of increase of the cavity volume for the cavitator moving at a constant velocity V_∞ is

$$\frac{dW}{dt} = Q_n = \pi \cdot R_x^2 \cdot V_\infty = \frac{\pi \cdot x}{\lambda^2}(L_c - x) \cdot V_\infty \tag{10}$$

In view of the expression for the supercavity length L_c, the law of variation of the injection rate Q_{in} can be written as

$$Q_{in}(x) = \frac{\pi \cdot x \cdot \sigma \cdot V_\infty}{\ln \frac{1}{\sigma}}\left(\frac{2 \cdot R_n}{\sigma}\sqrt{C_x \cdot \ln \frac{1}{\sigma}} - x\right). \tag{11}$$

The maximum gas injection rate will correspond to the maximum cross-section of the supercavity ($x = L_c/2$, $C_x = C_{x0}(1 + \sigma)$)

$$Q_{in\max} = \pi \cdot R_c^2 \cdot V_\infty = \pi \cdot R_n^2 V_\infty C_{x0} \frac{1+\sigma}{\sigma}. \tag{12}$$

The corresponding maximum and average injection coefficients will be

$$\bar{Q}_{in\max} = \frac{Q_{in\ \max}}{\pi \cdot R_n^2 \cdot V_\infty} = C_{x0}\frac{1+\sigma}{\sigma}$$

and (13)

$$\bar{Q}_{n\ mean} = \frac{W_c \cdot V_\infty}{L_{c0}\pi \cdot R_n^2 \cdot V_\infty} = \frac{2}{3}C_{x0}\frac{1+\sigma}{\sigma}.$$

The shortest possible time of development of the supercavity to its estimated length $L_c = L_{cO}$ can be assessed as

$$t_{\min} = \frac{L_{c0}}{V_\infty} = \frac{2 \cdot R_n}{\sigma \cdot V_\infty}\sqrt{C_{x0}(1+\sigma) \cdot \ln\frac{1}{\sigma}}. \tag{14}$$

For a disc cavitator $C_x = C_{x0}(1+\sigma)$ where $C_{x0} = 0.82$ is drag coefficient at $\sigma = 0$. At $\sigma = 0.02 \div 0.1$ $\bar{Q}_{in\max} \approx 42 \div 9$ [1].

It is significant that the gas entrainment coefficient for the same cavitation number range will be an order of magnitude smaller $\bar{Q}_{out} = 4 \div 0.5$ $(\sigma = 0.02 \div 0.1)$, [3].

5 Steady Motion Regime

For a steady motion $\dot{W} = 0$, and the required gas injection rate will only be limited by the cavity gas entrainment rate Q_{out}.

A water tunnel experiment on free axisymmetric cavities downstream of a disc has shown that the steady flow inside a cavity consists of two regions: a through flow region F_P and a circulation flow region F_R (Fig. 2).

The circulation flow inside a ventilated supercavity downstream of a disc cavitator was observed in water tunnel experiments conducted at the Institute of Hydromechanics of the National Academy of Sciences of Ukraine. Visualization was made using aluminium powder added to the gas being injected. In the

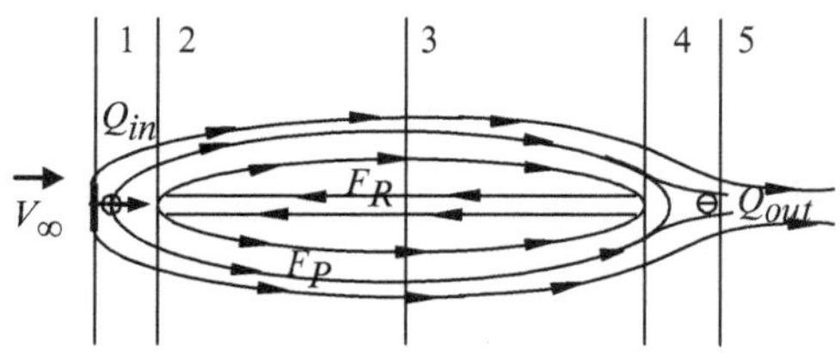

Fig. 2 Scheme of gas flow inside the cavity

experiments, a pressure difference of about 10% of the hydrostatic pressure between the leading and the trailing part of a supercavity was detected.

The through flow region F_p is formed by a source (+) and a sink (−) of the same intensity according to (7), which are situated distance L_c apart at the end points on the longitudinal axis of the elliptical supercavity. The circulation flow region F_R has the form of an ring vortex situated inside the elliptical supercavity. The steady circulation flow inside the vortex is sustained due to viscous interaction with the movable liquid boundary of the supercavity.

The velocity V_o in the axial flow of the vortex is in opposition to the inflow and equal to the liquid velocity $V_c$on the supercavity boundary. Because of this, the relative velocity V_e is twice the inflow velocity

$$V_e = V_o + V_c \approx 2V_\infty.$$

Conventionally, in the flow pattern in Fig. 2 five characteristic sections can be distinguished along the length of the supercavity:

Section S_1 is gas injection region. At this section, a gas is injected into the supercavity at flow rate Q_1 and relative velocity V_e

$$Q_1 = Q_{in}, V_e = V_\infty - Q_{in}/\pi R_n{}^2. \tag{15}$$

Section S_2 is the start of the circulation flow. It also gives the limit dimension of a supercavity with section F_2 where the injection rate Q_{in} is still sufficient for the supercavity cross-sections to expand. Cross-section F_2 is determined by the conditions (10) and (15):

$$Q_{in} = F_2 \cdot V_\infty = \pi \cdot R_2^2 \cdot V_\infty = \sigma \cdot \pi \cdot V_\infty \frac{x_2(L_C - x_2)}{\ln \frac{1}{\sigma}} \tag{16}$$

The region $S_2 - S_3$ is the region of expanding circulation flow.

Section S_3 is the supercavity midsection where the through flow reaches its minimum width

$$h = \frac{Q_{in}}{\pi \cdot D_c \cdot V_\infty}, \tag{17}$$

while the circulation flow reaches its maximum cross-sectional area

$$F_R = \frac{\pi(D_c - 2h)^2}{4}. \tag{18}$$

The region $S_3 - S_4$ is the cavity closure region where a re-entrant flow is formed in the central part of the circulation flow.

Section S_4 is the end of the circulation flow.

Section S_5 is the end of the supercavity and the start of its hydrodynamic wake where the volume of the gas determines its entrainment rate Q_{out} from the cavity. In a steady flow where the velocity, depth, injection rate, and cavitation number are constant, the gas entrainment rate is equal to the injection rate (7).

6 Gas Entrainment

The 0.001 s supercavity surface photos show the presence of sizeable disturbances (Fig. 3). Radial disturbances increase toward the trailing part of the cavity, and their size is about 1/100 of the full cavity length L_c. A similar pattern of destruction is shown by free jets issuing from a nozzle into the air. As shown schematically in Fig. 4, the gas fills voids in free boundary defects outside of the undisturbed cavity contour, and an equivalent volume of liquid in the form of droplets finds itself inside the supercavity contour.

It is quite natural to assume that on cavity closure the gas that has filled the defects outside of the disturbed boundary will remain in the cavity wake and the splashes will form foam, which will also be entrained into the wake, but by another entrainment mechanism.

The assumptions made allow one to estimate the gas entrainment rate from the gas content in the vicinity of the supercavity midsection

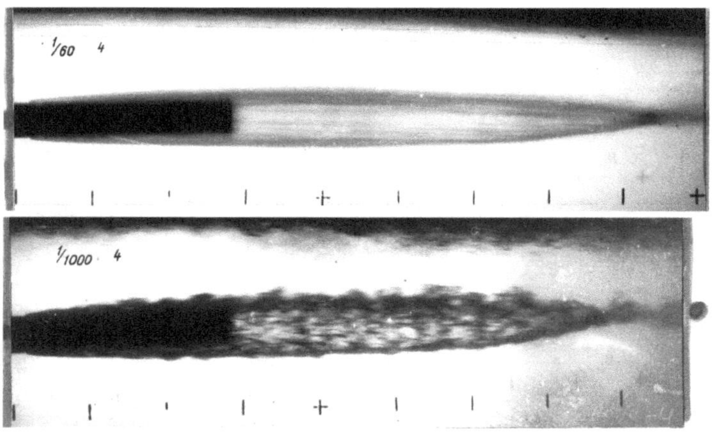

Fig. 3 Supercavity surface photos with different exposition time (1/60 s and 1/1000 s)

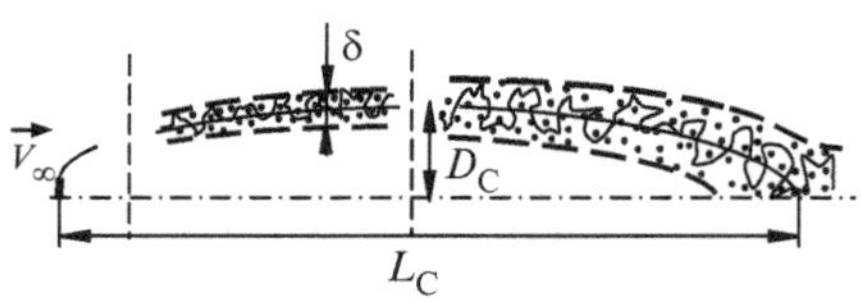

Fig. 4 Scheme of the disturbed supercavity surface

$$Q_{out} = K \cdot \pi \cdot D_c \cdot V_\infty \cdot \delta(\mathrm{Re}_L; \mathrm{We}; \mathrm{Fr}; \mathrm{Ri}), \tag{19}$$

where K is the gas concentration in the layer, $\pi \,.\, D_c \,.\, \delta$ is the cross-sectional area of the disturbed outer layer of the liquid supercavity boundary, δ ($\mathrm{Re_L}$; We; Fr; Ri) is the disturbed layer thickness, and $\mathrm{Ri} = \rho/\rho_g$ is the Richardson number.

If we neglect the Richardson number since in this case $\rho_g = \mathrm{const}$ and $\mathrm{Ri} = \mathrm{const}$, the Froude number $Fr = \frac{V_\infty}{\sqrt{g \cdot D_n}} > 10^2$, and the Weber number $We = \frac{\rho \cdot V_\infty^2 D_n}{\xi} > 10^3$, which are large enough, then δ will be a function of the Reynolds number $\mathrm{Re}_L = \frac{V_e L_c}{\nu}$ alone. Assuming that the disturbed layer thickness function has a structure similar to that for a boundary layer [15], δ will take the form

$$\delta = K_1 \cdot L_c \cdot (\mathrm{Re_L})^n, \tag{20}$$

where K_1 and n are to be determined for the conditions on the free cavity boundary.

The gas entrainment rate Q_{out} and the gas entrainment coefficient $\bar{Q}_{out}$ will be

$$Q_{out} = K \cdot K_1 \pi \cdot D_c L_c V_\infty (\mathrm{Re}_L)^n, \tag{21}$$

$$\bar{Q}_{out} = \frac{Q_{out}}{\pi \cdot R_n^2 V_\infty} = 4K \cdot K_1 C_{x0} \frac{1+\sigma}{\sigma} \sqrt{\frac{1}{\sigma} \ln \frac{1}{\sigma}} \cdot \mathrm{Re}_L^n, \tag{22}$$

$$\mathrm{Re}_L = \frac{V_e L_c}{2 \cdot \nu} = \left(V_\infty \sqrt{1+\sigma} - V_g\right) \frac{L_c}{2 \cdot \nu},$$

$$V_e = V_c - V_g = \left(V_\infty \sqrt{1+\sigma} - V_g\right).$$

Here, $C_{x0} = 0.82$ for a disc, Re_L is the Reynolds number based on the cavity length, and V_e is the relative velocity.

It is easy to see that the form of Eqs. 20–22 opens up new possibilities for varying the gas entrainment rate by varying the relative velocity in the number $\mathrm{Re_L}$.

Assuming that the layer thickness in Eq. 20 depends solely on the cavity length $\delta = \delta(L_c)$, the gas entrainment rate (22) will be a function of the cavitation number alone

$$Q_{out}(\sigma) = \mathrm{Const} \cdot \frac{1+\sigma}{\sigma} \sqrt{\frac{1}{\sigma} \ln \frac{1}{\sigma}}. \tag{23}$$

An equation of the form of (23) has also been derived using dimensional theory alone [19], thus confirming its fundamentality for the gas entrainment process.

7 A Solid Body in a Supercavity

A solid body in a supervcavity may affect gas entrainment by changing the supercavity volume and the gas flow velocity in the gap between the free boundary and the body surface.

If the body contour R_b and volume W_b are known

$$R_b = F_b(x), \quad W_b = \pi \int_0^{L_b} F_b^2(x)dx, \tag{24}$$

then the gas balance equation (5) will read

$$\frac{d(W - W_b)}{dt} = Q_{in} - Q_{out}. \tag{25}$$

The average flow velocity V_g in the gap $h(x) = R_x - R_b$ will be

$$V_g(x) = \frac{Q_{in}}{\pi \cdot h(R_x + R_b)}, \tag{26}$$

and the cavity boundary velocity will be [2, 3]

$$V_c = V_\infty \sqrt{1 + \sigma_n}. \tag{27}$$

In the case of a ventilated wall supercavity, the gas will be injected thereto from the solid wall using special nozzles (Fig. 5). In this case, the velocity gradient on the free boundary may be both positive and negative, and the relative velocity $V_e = V_c - V_g$ may reverse sign depending on the gap width h and the gas entrainment rate $Q_{out} = Q_{in}$.

$$V_e = V_c - V_g = V_\infty \sqrt{1 + \sigma} - \frac{Q_{in}}{\pi \cdot h(R_x - R_b)}. \tag{28}$$

Writing the condition for the absence of disturbances on the supercavity boundary (4) in the form

$$|V_e| < V_{gcr} \tag{29}$$

makes it possible to obtain the relationship for choosing optimum values of the gap h and the injection rate Q_{in}

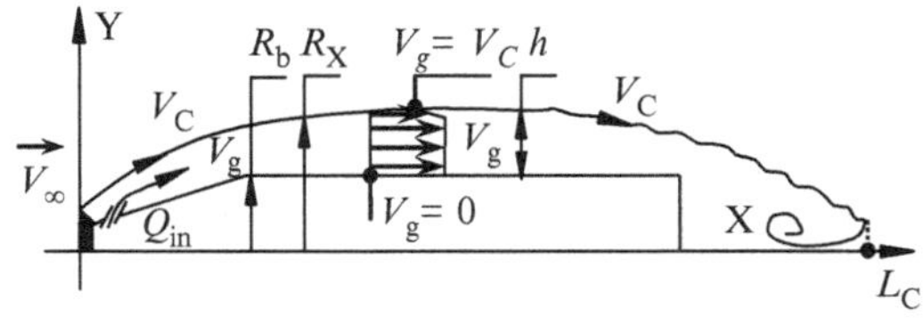

Fig. 5 The gas flow pattern in the gap between the supercavity surface and the solid body

$$\left|\pi(R_x + R_b)\left(V_\infty\sqrt{1+\sigma} - V_{gcr}\right)\right| < \frac{Q_{in}}{h}, \tag{30}$$

where $V_{gcr} = 13$ m/s.

Conclusions

It is suggested that Helmholtz–Taylor-type mechanisms of instability of the free boundary of a supercavity under the action of a gas flow be taken into account when considering supercavity boundary disturbances and calculating the gas entrainment rate.

Special gas injection regimes in the formation of a supercavity and for a steady supercavity are pointed out. Formulas are given to estimate the required gas injection rate.

A circulation flow and a through flow inside a supercavity are pointed out. Photos of disturbed supercavity boundaries are presented.

It is shown that there can exist near-wall gas flows with both a positive and a negative velocity gradient in the gap between the solid body boundary and the supercavity boundary.

References

1. Egorov IG. Ventilated cavitation (in Russian). In: Egorov IG, Sadovnikov YuM, Isaev II. Leningrad: Sudostroenie; 1971. 284pp.
2. Logvinovich GV. Free-boundary flow hydrodynamics (in Russian). Kiev: Naukova Dumka; 1969.
3. Epshtein LA. Methods of dimensional theory and scaling in vessel hydromechanics problems (in Russian). Leningrad: Sudostroenie; 1970.
4. Savchenko YuN, Semenenko VN, Putilin SI. Nonstationary processes in the motion of supercavitating bodies (in Russian). Prykladna Gidromekhanika. 1999;1(1):62–80.
5. Savchenko YuN, Semenenko VN, Putilin SI, Savchenko G, Naumova E. Designing the high-speed supercavitating vehicles. Proceedings of the 8th International Conference on Fast Sea Transportation (FAST'2005), St. Petersburg; June 2005. p. 1–7.
6. Cox PN, Claiden WA. Air entrainment of the rear of a steady cavity. Cavitation in Hydrodynamics. Proceedings of the Symposium, London; 1955.
7. Krylov VV. Experimental data on air entrainment from a cavity formed by air injection (in Russian). Moscow: TsAGI Transactions; 1961. Issue 824, 284pp.
8. Epshtein LA. On the mechanism of pulsation processes in the trailing part of attached cavities (in Russian). Proceedings of the Symposium on the Physics of Acustic/Hydrodynamic Phenomena. Moscow: Nauka; 1975. p. 133–8.

9. Spurk IH. A theory for the gas loss from ventilated cavities. Proceedings of the International Science School "High Speed Hydromechanics". Cheboksary; 2002. p. 191–5.
10. Spurk IH. On the gas loss of ventilated supercavities. Acta Mech. 2002;155(3–4):125–35.
11. Semenenko VN. Instability and oscillation of gas-filled supercavities. Proceedings of the Third International Symposium on Cavitation, Grenoble; April 1998. p. 25–30.
12. Franc J-P, Michel J-M. Fundamentals of cavitation. Dordrecht/Boston/London: Kluwer; 2004.
13. Ariel' NZ, Bortkovsky RS. Refined model of the energy and mass exchange of splashes over a storming ocean surface (in Russian). In: Typhoon-75 (Expedition materials). Leningrad; 1978. Vol. 2, p. 101–15.
14. Bortkovsky RS. Atmosphere–ocean heat and water exchange in a storm (in Russian). Leningrad: Gidrometeoizdat; 1983.
15. Preobrazhensky LYu. Estimation of the droplet and splash content in the surface layer of the atmosphere (in Russian). Trudy GGO; 1972. Issue 282, p. 194–9.
16. Savchenko GYu. Hydrodynamics of wall supercavity flows (in Russian). Kiev: PhD thesis; 2009. 138pp.
17. Lane WR, Green HL. The mechanics of drops and bubbles. In: Batchelor GK, Davies RM, editors. Surveys in mechanics. London/New York: Cambridge University Press; 1956. p. 162–215.
18. Birkhoff G. Hydrodynamics, methods, facts and similarity (in Russian). Moscow: Izdatelstvo Inostrannoi Literatury; 1963.
19. Mamenko YuN. Determination of conditions for the onset of splashing on a liquid surface acted upon by a plane gas jet (in Russian). Transactions of Kaliningrad Technological Institute of Fish Industry, Kaliningrad, USSR; 1980. Issue 90, p. 109–12.
20. Savchenko YuN, Semenenko VN. Wave generation on the boundaries of supercavities formed in the water entry of a disc and cones (in Russian). Problems in high-speed hydrodynamics (Transactions). Cheboksary: Chuvash University; 1993. p. 231–9.

Generalized Self-Similar Problem of Impact Between Liquid and Solid Wedges

Yu. N. Savchenko and Yu. A. Semenov

Abstract
This paper presents a generalized solution of the self-similar problem of impact between solid and liquid wedges, which includes as a special case the problem of asymmetric wedge entry into a liquid. The solution method is based on the construction of an analytical expression for the complex flow potential in a parameter region. From the dynamic and the kinematic boundary condition, integral equations are obtained for the determination of the velocity magnitude and angle with the free surface, which appear in the expression for the complex potential. The free boundary shape, the pressure distribution along the wedge, and the hydrodynamic force coefficients are calculated over a wide range of given data.

1 Introduction

Problems of body entry into a liquid are of great importance in such applications as the study of ship roll and pitch and offshore platform behavior in a heavy sea and the design of planing boats, semisubmerged propellers, and seaplanes.

The initial stage of body-liquid interaction is of a shock nature, and it is characterized by a high load level and substantial interplay between nonlinear and unsteady effects. This class of problems involves additional difficulties, which are due to the presence of a free boundary and a three-phase contact line at the body-liquid-air interface. The body-liquid entry problem is overviewed in considerable detail in Korobkin and Pukhnachev [1]. By now, only the symmetric entry of a wedge into a liquid with an unperturbed free surface at a constant velocity has been studied in detail.

Yu.A. Semenov (✉)
Institute of Hydromechanics of the National Academy of Sciences of Ukraine, Kyiv, Ukraine
e-mail: semenov@a-teleport.com

I. Nesteruk (ed.), *Supercavitation*,
DOI 10.1007/978-3-642-23656-3_8, © Springer-Verlag Berlin Heidelberg 2012

In actual practice, the water surface interacts with the atmosphere, as a result of which its shape becomes wavy. In this connection, a more accurate evaluation of loads arising from hydroplaning, seaplane landing, and offshore platform – heavy sea interaction calls for taking into account the actual shape of the free surface. As known from wave theory, the limiting wave crest form is a 120° wedge. The model of solid body – liquid wedge interaction presented in this paper allows one to evaluate the effect of the wedge shape of a liquid on the unsteady hydrodynamic loads arising from impact on a solid body. In this model, a flat free surface is a special case.

Karman [2] and Wagner [3] were the first who obtained an approximate solution of the entry of a wedge into a flat-surface liquid at small deadrise angles. They considered the water entry process as a sequence of pulses generated by the body-liquid impact. The study was continued, in particular by Garabedian [4], Borg [5], Logvinovich [6], and Moiseev [7]. Gonor [8] considered the entry of a thin wedge. Mackie [9] obtained a complete linearized solution. Logvinovich [6] showed, among other things, that as the deadrise angle approaches zero, the liquid speed in the tip jets becomes twice that in the flow turn and tip jet formation region.

Dobrovol'skaya [10] obtained a complete solution to the nonlinear problem of the vertical entry of a symmetric wedge by constructing the Wagner function. She found the Wagner function by conformal mapping and formulated a boundary-value problem for the mapping function in such a way as to find it using Schwartz's integral formula. Dobrovol'skaya derived a singular integral equation in the real part of the mapping function on the free surface and presented numerical results for wedge angles less than 30°. A numerical solution of the singular integral equation for larger wedge angles was given by Zhao and Faltinsen [11], Keady and Fowkes [12], and Freankel [13, 14].

Logvinovich [6] set forth a theory of the entry of finite-size bodies, for which Wagner's assumptions do not hold, and his theoretical estimates were verified by experiment. Since the forces that arise from high-speed water entry may be far greater than their steady values, liquid compressibility and body elasticity may play a significant role. These issues were considered by Sagomonian [15], Troshin et al. [16], Kubenko [17], and Korobkin [18].

Cavitation in axisymmetric body entry was simulated using Logvinovich's semiempirical theory by Zhuravlev [19] and then by Savchenko and Semenenko [20, 21]. They found and explained the interesting effect of wave generation on cavities, which is due to oscillations of the air entrained by the cavity from the atmosphere.

The application of numerical methods (the finite-element and the boundary-element method) to water entry problems presents difficulties because of the salient points of the flow boundary at the three-phase contacts. At these points, the flow potential has singularities, thus making it difficult to obtain reliable numerical results. As a consequence, in [11, 22] the tip jet flow region was truncated, and in [21] an asymptotic analytical solution was used for the tip jet flow region. The numerical results for the entry of an asymmetric wedge reported in the literature are limited to small angles between the wedge axis and the vertical, at which effects of asymmetric interaction between the wedge sides and the liquid are rather insignificant.

This paper presents a generalized solution of the problem of impact between a solid and a liquid wedge. A liquid wedge angle of 180° corresponds to an unperturbed flat surface. This case for the vertical and the oblique entry of a wedge into an unperturbed liquid has been considered in [23–26].

This paper also discusses the limit combination of the horizontal component of the entry velocity and the wedge orientation at which flow separation occurs and only one wedge side interacts with the liquid [27]. The solution method is based on the construction of an analytical expression for the complex flow potential in a parameter region. From the dynamic and the kinematic boundary condition, integral equations are obtained for the determination of the velocity magnitude and angle with the free surface, which appear in the expression for the complex potential. The free boundary shape and hydrodynamic force coefficients are calculated.

2 Formulation of the Boundary-Value Problem of Impact Between Liquid and Solid Wedges

As distinct from the extensive literature on the vertical entry of a symmetric wedge into a liquid, few works are devoted to a more general problem – the vertical entry of an asymmetric wedge. Garabedian [4] and Borg [5] were the first to attack this problem. More recent publications on the subject are Chekin [28] and Korobkin [29].

Chekin's method is based on the use of Sokhotski's integral formula to find a function that maps the upper half-plane onto a stationary flow plane in self-similar variables. Chekin's solution for asymmetric wedge entry assumes flow separation at the wedge vertex followed by flow detachment to a wedge side. However, the experiments by Judge and others [30, 31] have shown that this assumption does not hold. Only in the special case where the stagnation point is at the wedge vertex, Chekin's solution describes nonseparated flow.

In [32], an attempt was made to estimate the hydrodynamic flow characteristics for the entry of an asymmetric wedge by constructing an approximate physical and mathematical model, which includes an approximate expression for the complex flow potential. However, the calculated results differ considerably from the experimental ones [30, 31].

In this paper, a similar approach is used, and the free boundary of the liquid is assumed to be in the form of a wedge. The velocity at infinity has a magnitude V_∞ and makes an angle γ_∞ with the symmetry axis of the liquid wedge; the bisector of the solid wedge makes an angle δ with the velocity direction. The angle δ is positive if the wedge is rotated counterclockwise from the velocity vector. Let 2α be the wedge angle; then $\beta_L = \pi - \alpha - \delta - \gamma_\infty$ and $\beta_R = \gamma_\infty - \alpha + \delta$ are the angles of the left and the right wedge sides with the horizontal. For a constant entry velocity, the introduction of the self-similar variables $x = X/V_0t$ and $y = Y/V_0t$ where t is the time allows one to transform the time- In [23], a solution for reversible flow is

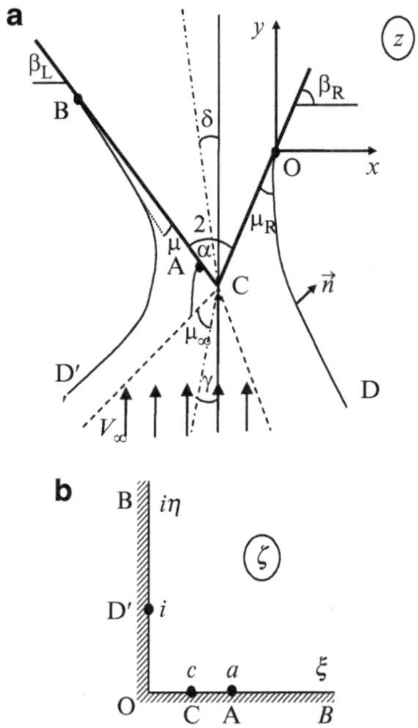

Fig. 1 Physical flow plane (**a**) and parameter region (**b**)

sought for. A fixed wedge in an unbounded flow is considered. varying flow boundary in the physical plane $Z = X + iY$ into a stationary boundary in the plane $z = x + iy$ where V_0 is the velocity magnitude at the point of contact of the free surface with the right wedge side (Fig. 1). Since V_0 is chosen as the characteristic flow speed, the dimensionless speed in the plane $z = x + iy$ at point O is equal to unity ($v_O = 1$), [23].

The complex potential of the self-similar flow $W(Z,t) = \Phi(Z,t) + i\Psi(Z,t)$ can be represented as

$$W(Z,t) = V_0^2 t w(z) = V_0^2 t[\varphi(z) + i\psi(z)], \tag{1}$$

where φ and ψ are the velocity potential and the stream function in the stationary plane z. The function $w(z)$ is to be found.

Following Joukovski [33], the solution is sought for by constructing two functions: the complex velocity, dw/dz, and the derivative of the complex potential, $dw/d\varsigma$, in the region of a parametric variable, $\varsigma = \xi + i\eta$, for which the first quadrant is chosen. Conformal mapping allows us to fix three arbitrary points in the parameter region. If such points are O, B, and D (Fig. 1), then the wetted surface of the wedge and the free surface will correspond to the real and the imaginary axis, respectively. The stagnation point A and the wedge vertex C correspond to the points $\varsigma = a$ and $\varsigma = c$ in the parameter plane, which are to be found as part of the solution.

If the functions dw/dz и $dw/d\varsigma$ are found, the velocity field in the parameter plane is known, and the parameter plane and the stationary plane z are related to each other as

$$v_x - iv_y = \frac{dw}{dz}(\varsigma), \quad z(\varsigma) = z(0) + \int_0^{\varsigma} \frac{dw}{d\varsigma}\frac{dw}{dz} d\varsigma, \tag{2}$$

where v_x and v_y are the x- and y-components of the velocity.

2.1 Complex Velocity

The boundary-value problem for the complex velocity function can be formulated as follows. For now, let us assume that the velocity modulus along the free surface, i.e. along the imaginary axis of the first quadrant is known

$$v(\eta) = \left|\frac{dw}{dz}\right|, \qquad 0<\eta<\infty, \quad \xi = 0, \tag{3}$$

In the wedge-bound coordinate system, the normal velocity component on the wedge faces is zero due to the impermeability condition. This means that the argument of the complex velocity along the real axis of the first quadrant is fixed and determined by the wedge orientation.

$$\chi(\xi) = arg\left(\frac{dw}{dz}\right) = \begin{cases} -\beta_R, & 0<\xi<a, \ \eta = 0, \\ -\pi - \beta_R, a<\xi<c, & \eta = 0, \\ -\pi + \beta_L, c<\xi<\infty, & \eta = 0. \end{cases} \tag{4}$$

The integral formula [34]

$$\frac{dw}{dz} = v_\infty \exp\left[\frac{1}{\pi}\int_0^\infty \frac{d\chi}{d\xi'} \ln\left(\frac{\varsigma + \xi'}{\varsigma - \xi'}\right) d\xi' - \frac{i}{\pi}\int_0^\infty \frac{d\ln v}{d\eta'} \ln\left(\frac{\varsigma - \eta'}{\varsigma + \eta'}\right) d\eta' + i\chi_\infty\right] \tag{5}$$

makes it possible to solve the boundary-value problem (3) and (4) in the first quadrant of the complex plane ς and find the complex velocity as

$$\frac{dw}{dz} = e^{-i(\gamma_\infty+\alpha+\delta)}\left(\frac{\varsigma - a}{\varsigma + a}\right)\left(\frac{\varsigma + c}{\varsigma - c}\right)^{(1-2\alpha/\pi)}$$
$$+ \, exp\left[-\frac{i}{\pi}\int_0^\infty \frac{d\ln v}{d\eta}\ln\left(\frac{i\eta - \varsigma}{i\eta + \varsigma}\right)d\eta\right]. \quad (6)$$

Substituting $\varsigma = \xi$ into Eq. 6 shows that $\arg(dw/dz)$ satisfies the boundary condition (4), and at $\varsigma = i\eta$ the complex velocity satisfies the condition (3).

2.2 Derivative of the Complex Potential

To analyze the behavior of the velocity potential along the free boundary, it is convenient to introduce two unit vectors, $\mathbf{n}$ and $\boldsymbol{\tau}$, which are normal and tangent to the free boundary, respectively. The normal unit vector points outward from the liquid, and the tangent vector is directed so that the arc coordinate s increases in the direction for which the liquid is on the left (Fig. 1). With this notation

$$dw = (v_s + iv_n)ds, \quad (7)$$

where v_s and v_n are the tangential and normal velocity components, respectively. Let θ be the angle between the velocity on the free boundary and the unit vector τ: $\theta = \tan^{-1}(v_n/v_s)$. The behavior of the function θ along the liquid boundary is shown in Fig. 2.

Equation 7 gives the argument of the derivative $dw/d\varsigma$ of the complex potential

$$\vartheta(\varsigma) = \arg\left(\frac{dw}{d\varsigma}\right) = \arg\left(\frac{dw}{ds}\right) + \arg\left(\frac{ds}{d\varsigma}\right)$$
$$= \begin{cases} \theta, & 0<\xi<\infty, \quad \eta = 0, \\ \theta - \pi/2, & \xi = 0, \quad 0<\eta<\infty. \end{cases} \quad (8)$$

Consider the behavior of the function $\theta(\varsigma)$ along the whole of the liquid boundary, i.e. along the real and the imaginary axis of the parameter region. When moving along the free boundary from point O to point D, the function $\theta(\varsigma)$ increases from the value μ_R at $\varsigma = 0$ to the value $\mu_\infty + \gamma$, which corresponds to the velocity direction at infinity (the point $\varsigma = i$). To reach the left infinite point of the free boundary where the function θ takes the value $\mu_\infty - \gamma$ with the velocity direction unchanged and equal to its value at infinity, we have to move along a closed contour of infinite radius. To this contour there corresponds a circular arc of infinitesimal radius centered at the point $\varsigma = i$. On going around this semicircle, the function $\theta(\varsigma)$ is incremented by $\Delta\theta_D = -(\pi + 2\mu_\infty)$. The continuous variation of the function $\theta(\varsigma)$ is shown in Fig. 2 as solid lines, and its step changes are shown as dashed lines.

Fig. 2 Domain of variation of the function $\theta(\eta)$

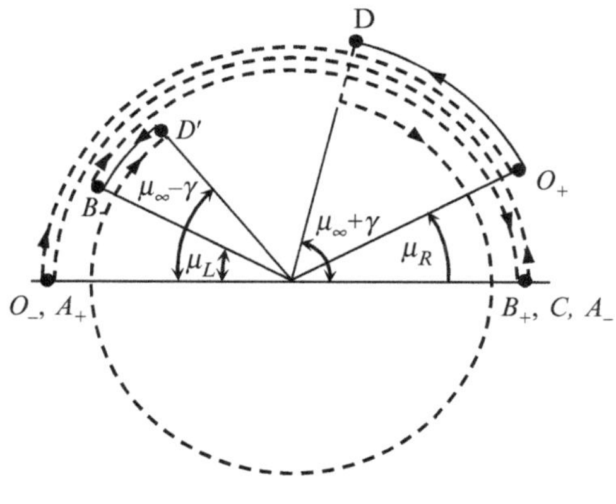

Then the function $\theta(\varsigma)$ varies continuously when moving along the free boundary from point D' to point B. On the interval $a<\xi<\infty, \eta=0$, which corresponds to the left side of the wedge and the part of the right side between points C and A, the function $\theta(\varsigma)\equiv 0$ since $v_n=0$ and $v_s>0$. On the interval $0<\xi<a,\ \eta=0$ the function $\theta(\varsigma)\equiv\pi$ since $v_n=0$ and $v_s<0$. Thus at the point $\varsigma=a$ the function $\theta(\varsigma)$ undergoes the step change $\Delta\theta_A=-\pi$. One more step change, $\Delta\theta_O=\mu_R-\pi$, takes place at point O on going in an infinitesimal neighborhood of the point $\varsigma=0$ from the wedge side, $\xi>0,\ \ \eta=0$, to the free surface, $\xi=0,\ \ \eta>0$. The function $\theta(\varsigma)$ can be expressed in terms of the continuous function $\lambda(\varsigma)$ defined as follows:

$$\theta(\varsigma)=\begin{cases}\lambda(\varsigma), & \xi=0,\quad 0<\eta<1,\\ \lambda(\varsigma)+\Delta\theta_D, & \xi=0,\quad 1<\eta<\infty,\\ \lambda(\varsigma)+\Delta\theta_O, & 0<\xi<a,\quad \eta=0,\\ \lambda(\varsigma)+\Delta\theta_O+\Delta\theta_A, a<\xi<\infty, & \eta=0,\end{cases} \tag{9}$$

where $\Delta\theta_D=-(\pi+2\mu_\infty)$ and $\Delta\theta_O=\mu_R-\pi$,

Equation 8 together with Eq. 9 specifies a homogeneous boundary-value problem for the function $dw/d\varsigma$. The function $dw/d\varsigma$ can be found using the integral formula [34]

$$\frac{dw}{d\varsigma}=K\exp\left[-\frac{1}{\pi}\int\limits_0^\infty\frac{d\chi}{d\xi'}\ln\left(\varsigma^2-\xi'^2\right)d\xi'\right] + \left[\frac{1}{\pi}\int\limits_0^\infty\frac{d\vartheta}{d\eta'}\ln\left(\varsigma^2+\eta'^2\right)d\eta'+i\chi_\infty\right], \tag{10}$$

which gives a solution of the boundary-value problem (8) in the first quadrant of the complex plane ς. Here, K is a real constant, and $\chi_\infty=\chi(\xi)|_{\xi=\infty}$.

Substituting Eqs. 8 and 9 into the first integral in (10) for ς varying along the real axis of the parameter region and into the second integral for ς varying along its imaginary axis and evaluating the integrals over the step changes of the function $\theta(\varsigma)$, we obtain the following expression for the derivative of the complex potential in the parameter plane

$$\frac{dw}{d\varsigma} = K\varsigma^{2\mu_R/\pi-1} \frac{\varsigma^2 - a^2}{(\varsigma^2+1)^{1+2\mu_\infty/\pi}} \times \exp\left[\frac{1}{\pi}\int_0^\infty \frac{d\lambda}{d\eta'} \ln\left(\varsigma^2 + \eta'^2\right) d\eta'\right]. \tag{11}$$

Integrating Eq. 7 in the parameter region gives the function that conformally maps the first quadrant onto the domain of variation of the complex potential

$$w(\varsigma) = w(0) + K\int_0^\varsigma \varsigma^{2\mu_R/\pi-1} \frac{(\varsigma^2 - a^2)}{(\varsigma^2+1)^{1+2\mu_\infty/\pi}} \times \exp\left[\frac{1}{\pi}\int_0^\infty \frac{d\lambda}{d\eta'} \ln\left(\varsigma^2 + \eta'^2\right) d\eta'\right] d\varsigma. \tag{12}$$

The velocity potential at point O, $w(0)$, can be put equal to zero. From Eqs. 6 and 11 we obtain the derivative of the mapping function

$$\frac{dz}{d\varsigma} = K\varsigma^{(2\mu_1/\pi-1)} \frac{(\varsigma + a)^2}{(1+\varsigma^2)^{1+2\mu_\infty/\pi}} \left(\frac{\varsigma - c}{\varsigma + c}\right)^{\left(1-\frac{2\alpha}{\pi}\right)} \times \exp\left[\frac{1}{\pi}\int_0^\infty \frac{d\lambda}{d\eta} \ln\left(\eta'^2 + \varsigma^2\right) d\eta' + i\gamma_\infty + \frac{i}{\pi}\int_0^\infty \frac{d\ln v}{d\eta'} \ln\left(\frac{i\eta' - \varsigma}{i\eta' + \varsigma}\right) d\eta' + i\delta + i\alpha\right], \tag{13}$$

whose integration in the parameter region gives the shape of the free surface.

The unknown parameters a, c, K can be found from the condition for the flow velocity at infinity and the condition for the length of the wetted parts of the wedge OC и CB.

2.3 Dynamic Boundary Condition

The Cauchy–Lagrange integral written in the physical region for point O and an arbitrary point in the flow has the form

$$\left.\frac{\partial \Phi}{\partial t}\right|_{z}+\frac{V^{2}}{2}+\frac{P}{\rho}=\left.\frac{\partial \Phi}{\partial t}\right|_{z=0}+\frac{V_{0}^{2}}{2}+\frac{P_{a}}{\rho}, \tag{14}$$

where P_a is the pressure on the free surface, which is assumed to be constant.

Let the free surface shape be described by the function $Z = Z(S, t)$ where S is the arc length along the free surface from point O, and t is the time. In [23], using the self-similar variable $s = S/(v_0 t)$, Eq. 10 is reduced to the differential equation, which relates the derivative of the velocity magnitude to the derivative of the velocity angle with the free boundary

$$\frac{dv}{ds}=\frac{v\,s\,\sin\theta}{v+s\cos\theta}\frac{d\theta}{ds}. \tag{15}$$

This equation is derived only in the assumption of flow self-similarity. Thus it holds for a variety of self-similar problems.

Multiplying Eq. 15 by $ds/d\eta$ gives the integro-differential equation

$$\frac{dv}{d\eta}=\frac{v\,s\,\sin\theta}{v+s\cos\theta}\frac{d\theta}{d\eta}, \tag{16}$$

where

$$s(\eta)=-\int_{0}^{\eta}\left|\frac{dz}{d\varsigma}\right|_{\varsigma=i\eta}d\eta=-K\int_{0}^{\eta}\frac{\eta^{2\mu_R/\pi-1}}{v(\eta)}\frac{\eta^{2}+a^{2}}{\left(1-\eta^{2}\right)^{1+2\mu_\infty/\pi}}$$
$$\times\exp\left[\frac{1}{\pi}\int_{0}^{\infty}\frac{d\theta}{d\eta'}\ln\left|\eta'^{2}-\eta^{2}\right|d\eta'\right]d\eta. \tag{17}$$

2.4 Kinematic Boundary Condition

On the free surface, the pressure is constant, and thus the acceleration of a liquid particle is normal to the free surface. This condition can be written as [35]

$$Re\left(\frac{\partial U}{\partial t}\overline{dZ}\right)=0. \tag{18}$$

In [23], using self-similar variables, the following integral equation in the function $d\ln v/d\eta$ is obtained, which also holds in the case of impact between a solid and a liquid wedge

$$-\frac{1}{2\tan\theta}\frac{d\ln v}{d\eta}+\frac{1}{\pi}\int_0^\infty \frac{d\ln v}{d\eta'}\frac{\eta'}{\eta'^2-\eta^2}d\eta' \\ =\frac{a}{a^2+\eta^2}+\left(\frac{2\alpha}{\pi}-1\right)\frac{c}{c^2+\eta^2}. \tag{19}$$

The system of Eqs. 16 and 19 is closed and allows one to find the functions $v(\eta)$ and $\theta(\eta)$.

On the wedge surface, the normal velocity component is zero, and hence $\theta_R=\pi$ at point O and $\theta_L=0$ at point B. At the same points on the free surface, the function $\theta(\eta)$ is determined from the solution of the total system of equations. Thus the contact angles μ_R and μ_L of the free surface with the wedge sides are

$$\mu_R=\theta_R-\left[\pi-\lim_{\eta\to 0}\theta(\eta)\right]=\theta(0), \tag{20}$$

$$\mu_L=\left[\pi-\lim_{\eta\to\infty}\theta(\eta)\right]-\theta_L=\pi-\lim_{\eta\to\infty}\theta(\eta). \tag{21}$$

The Cauchy–Lagrange integral (10) allows us to calculate the pressure on the wedge sides. To do this, we have to express the derivative $\partial\Phi/\partial t$ of the flow potential in the physical region in terms of the potential ϕ in the stationary region; then the expressions for the pressure coefficient $c_p=(P-P_a)/(\rho V_\infty^2)$ become: on the left wedge side from the contact point O to the stagnation point A

$$c_{p1}(s)=-\frac{2(\phi+s\,v)+(1-v)^2}{v_\infty^2},\quad 0\le\xi\le a; \tag{22}$$

from the stagnation point on left wedge side to the wedge vertex and on the right wedge side

$$c_{p2}(s)=-\frac{2(\phi-s\,v)+(1+v)^2}{v_\infty^2},\quad a\le\xi\le\infty, \tag{23}$$

where ϕ, v, and s are determined from Eqs. 6, 12, and 13

$$\phi(\xi) = \Re\left(w(\varsigma)|_{\varsigma=\xi}\right), \quad v(\xi) = \left|\frac{dw}{dz}\right|_{\varsigma=\xi},$$

$$s(\xi) = \int_0^{\xi} \left|\frac{dz}{d\varsigma}\right|_{\varsigma=\xi'} d\xi'.$$

The hydrodynamic force coefficients of are obtained by integration of the pressure coefficient along the wetted part of the wedge according to the expressions

$$C_{nR} = \frac{1}{\frac{1}{2}\rho V_\infty^2 H} \int_0^a c_{p1}(s) \frac{ds}{d\xi} d\xi + \frac{1}{\frac{1}{2}\rho V_\infty^2 H} \int_a^c c_{p2}(s) \frac{ds}{d\xi} d\xi, \tag{24}$$

$$C_{nL} = \frac{1}{\frac{1}{2}\rho V_\infty^2 H} \int_c^\infty c_{p2}(s) \frac{ds}{d\xi} d\xi. \tag{25}$$

The section of the wedge by a straight line perpendicular to the velocity vector is chosen as the characteristic length

$$H = V_\infty t[(\alpha - \delta) + (\alpha + \delta)].$$

3 Calculated Results

3.1 Numerical Solution of the System of Equations

The system of the integral equations (16) and (19) is solved by the method of successive approximations, in which the function $d\ln v/d\eta$ is found from the integral equation (19) in the $(k+1)$-th iteration by taking the Hilbert transformation

$$\left(\frac{d\ln v}{d\eta}\right)^{(k+1)} = \frac{4}{\pi} \int_0^\infty \left\{ \frac{1}{2\tan\theta} \frac{d\ln v}{d\eta'} + \frac{a}{a^2 + \eta'^2} + \left(\frac{2\alpha}{\pi} - 1\right) \frac{c}{c^2 + \eta'^2} \right\}^k \frac{\eta'}{\eta'^2 - \eta^2} d\eta'. \tag{26}$$

From Eq. 16 we can find the $(k+1)$-th approximation for the derivative $d\theta/d\eta$,

$$\left(\frac{d\theta}{d\eta}\right)^{k+1} = \frac{v^{k+1} + s^k \cos\theta^k}{s^k \sin\theta^k} \left(\frac{d\ln v}{d\eta}\right)^{k+1}, \tag{27}$$

whose integration allows us to find the $(k+1)$-th approximation for the function $\theta(\eta)$. The functions $d\theta/d\eta$ and $\theta(\eta)$ have a singularity at point O as can be seen from Eqs. 26 and 27. On the other hand, these functions are bounded by their definition. To evaluate the integral in Eq. 26, the lower limit of integration is put equal to some small ε. As the initial approximation, we put $v(\eta) \equiv V_0$, $\theta(\eta) \equiv 0$.

In discrete form, the functions $d\theta/d\eta$ и $\theta(\eta)$ are approximated by piecewise constant functions on the intervals $0<\eta_i<1$, which correspond to the right portion of the free boundary. The fixed points η_i, $i = 1, N/2$ are distributed along the imaginary axis of the parameter region as a geometric series with the first term $\eta_1 = 10^{-4}$. On the intervals $1<\eta_i<\infty$, $i = N/2+1, N$, which correspond to the left portion of the free boundary, the points η_i are specified as $\eta_i = 1/\eta_{N-i}$. The length of the segment $s_1$nearest to point $O$is calculated analytically using Eq. 17

$$s_1 = -Ka^2 \exp\left(\frac{2}{\pi}\int_0^\infty \frac{d\theta}{d\eta'} \ln\eta' d\eta'\right) \frac{\eta_1^{2\mu_R/\pi}}{2\mu_R/\pi}. \tag{28}$$

The length of the segment s_N nearest to point B is determined in a similar way

$$s_N = \frac{K}{v_B} \frac{\eta_N^{-2\mu_L/\pi}}{2\mu_L/\pi}. \tag{29}$$

3.2 Impact Between Solid and Liquid Wedges

To verify the obtained theoretical results and to analyze the effect of the distribution of the points η_i and their number N on the calculation accuracy, the vertical entry of a wedge into a liquid was calculated. This case is considered in detail in [9–11, 13].

Figure 3 shows the calculated results for an impact between a solid and a liquid wedge, both of vertex angle 45°, which are compared with those obtained in [36] using the boundary-element method. In [38], the problem is formulated using self-similar variables. As illustrated, the calculated results are in reasonable agreement.

Figure 4 shows the profile of the free boundary and streamlines for an impact between a solid and a liquid wedge. The wetted part of the wedge side is also shown as a solid line, and its extension is shown as a dashed line. The case in Fig. 4b corresponds to a cumulative motion. In the tip jets, the liquid speed is several times as high as the upstream speed, and the horizontal velocity component is of opposite sign.

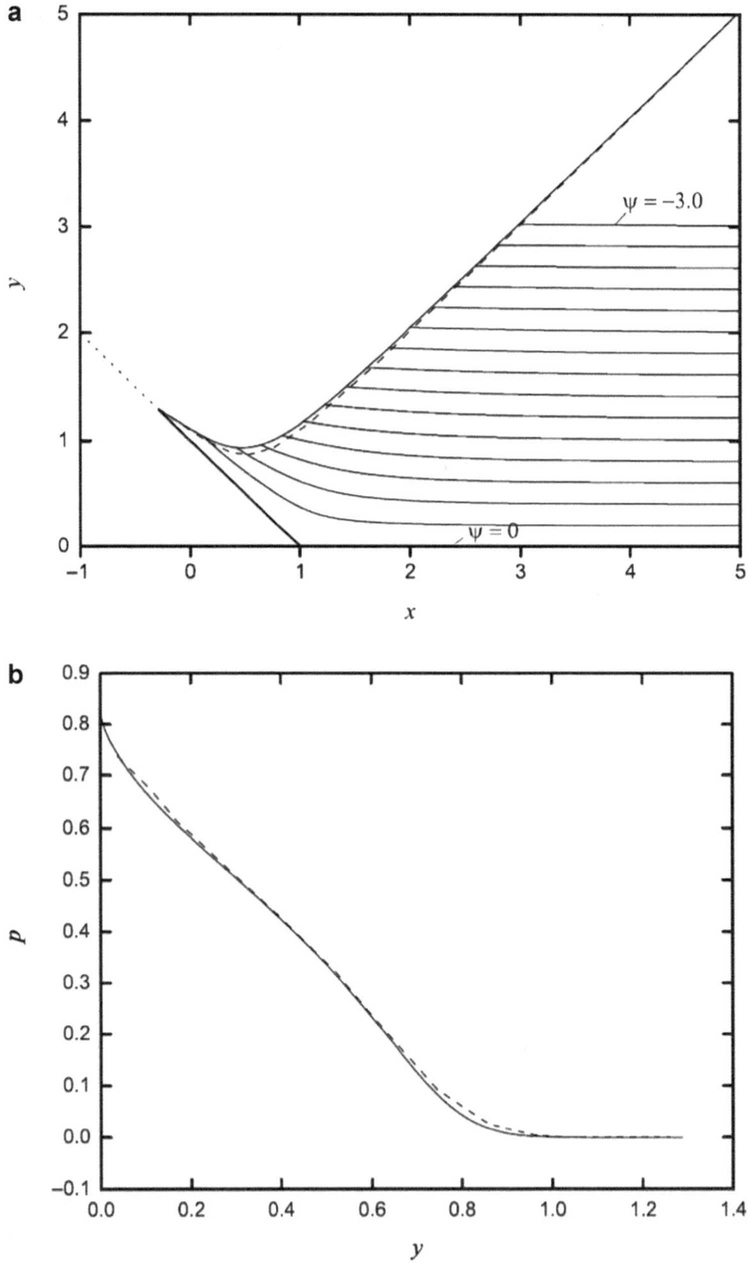

Fig. 3 Streamlines (**a**) and pressure distributions along the wedge sides (**b**) for solid and liquid wedge half-angles $\alpha = 45°$ and $\gamma = 45°$. The results obtained in [36, 37] using the boundary-element method are shown as *dashed lines*

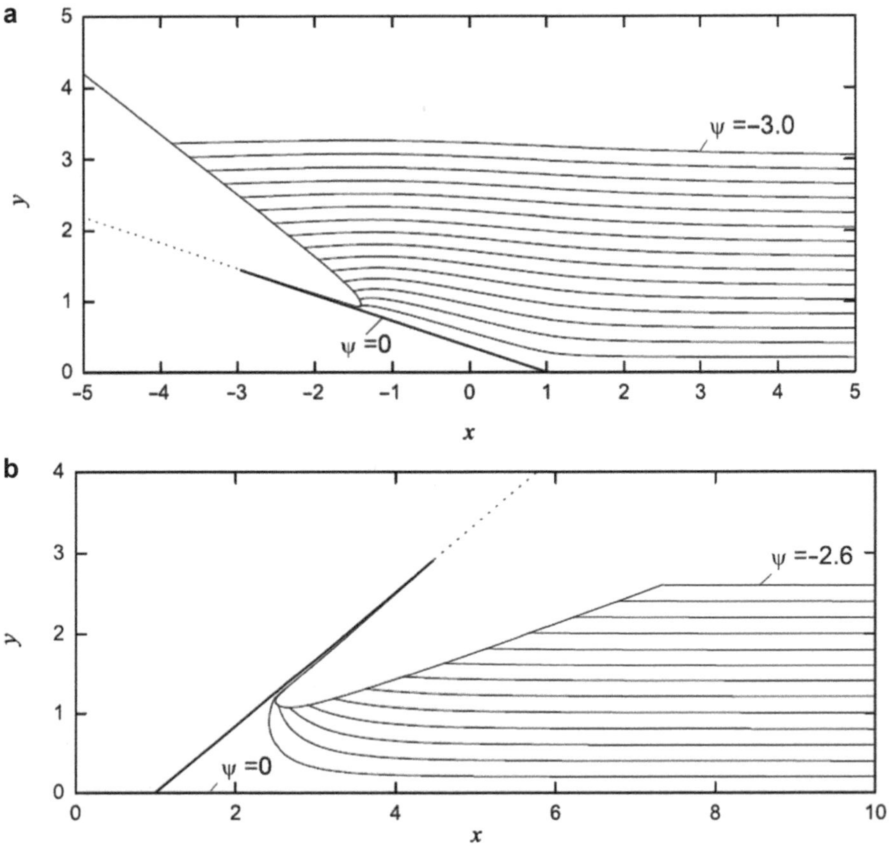

Fig. 4 Streamlines for an impact between a solid and a liquid wedge with half-angles $\alpha = 20°$, $\gamma = 140°$ (**a**) and $\alpha = 140°$, $\gamma = 20°$ (**b**)

The pressure distribution for the cases in Fig. 4 is shown in Fig. 5. One can see a local pressure increase in the tip jet root, which accelerates the liquid in the tip jet. This is the so-called slamming effect for the case in Fig. 4a or the cumulative effect for the case in Fig. 4b. In the latter case, as can be seen from Fig. 5, the pressure coefficient is several times as high as the pressure at the stagnation point for steady flows.

3.3 Limit Parameter Combination for Oblique Wedge Entry

The wedge orientation with respect to the free surface and the horizontal velocity component may be such that the liquid at infinity moves away from one of the wedge sides. With such a combination of given parameters, flow separation may occur at the wedge vertex, which changes the flow topology and calls for problem re-formulation. When flow separation occurs at the wedge vertex, only one wedge

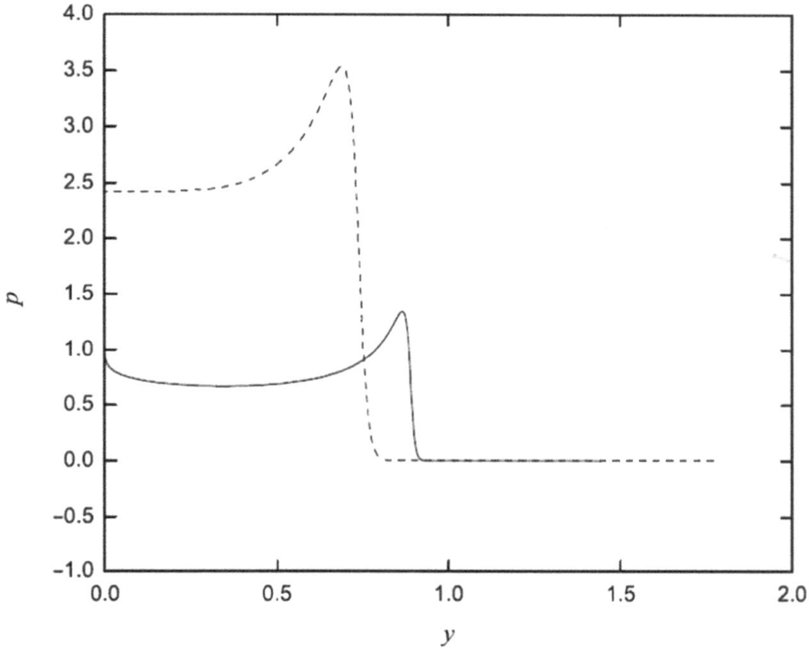

Fig. 5 Pressure distribution along the wedge sides for the cases in Figs. 4a (*solid line*) and Fig.4b (*dashed line*)

side interacts with the liquid, and the resulting flow corresponds to the entry of a plate.

The experimental and theoretical studies of flow separation and the formation of a ventilated cavity described in [30] made it possible to determine the angle of the wedge velocity at which the wedge flow changes to the plate flow. Based on the solution of the problem of the vertical entry of an asymmetric wedge, a flow separation criterion was proposed in [25]. The criterion depends on the determination of the normal force acting on a plate: according to it, flow separation occurs if the total force acting on one of the wedge sides becomes zero. The criterion is formulated without regard for the pressure on the free surface, which is justified for high wedge entry speeds. However, entry speeds may be moderate in such applications as slamming and the carrying capacity of high-speed planning boats.

If in the solution presented above the liquid wedge vertex angle is put equal to $\pi/2$, the flow will correspond to the oblique entry of a wedge into an unperturbed liquid whose flat surface is inclined at angle γ to the horizontal. Let Ω be the angle between the wedge symmetry axis and the perpendicular to the free surface and γ^* be the angle between the velocity and the free surface.

Figure 6 shows the flow configuration and streamlines corresponding to the limit angle γ^* for angles $\Omega = 0$ (symmetric entry with respect to the free surface) and $\Omega = 20°$. It can be seen that the streamlines on the left side of the wedge become nearly rectilinear and uniform.

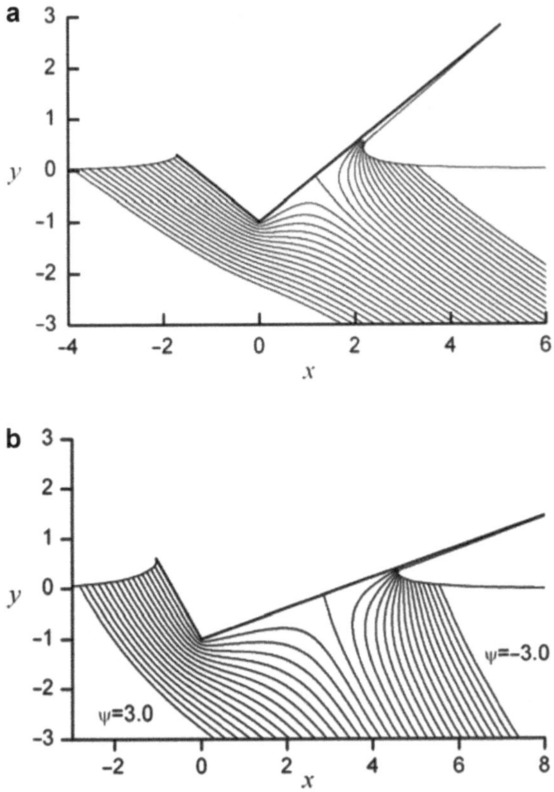

Fig. 6 Flow configuration corresponding to the onset of flow separation from the left side of a $2\alpha = 106°$wedge: (**a**) rotation $\Omega = 0°$, $\gamma^* = 59°$; (**b**) $\Omega = 20°$, $\gamma^* = 41°$

This means that the velocity in this portion of the free surface differs only slightly from the upstream velocity. There can be seen some elevation of the free surface, which is due to liquid inflow from the right.

The tip jet contact angles μ_R and μ_L versus the angle γ^* are shown in Fig. 7. The contact angle μ_R decreases and the contact angle μ_L increases as the angle γ^* increases to some limit value, which depends on the wedge orientation angle Ω. To the minimum angle γ^* at which the system of the integral equations (16) and (19) can be solved corresponds contact angle $\mu_L/\pi = 0.1$ for all wedge orientation angles in Fig. 8. This is in agreement with the results reported in [10–12] and [14] for the symmetric entry of a wedge. As $\alpha \to 0$, the contact angle $\mu/\pi \to 0.1$. Since for deadrise angles $\beta<\pi/2$ contact angles $\mu_L/\pi>0.1$ are nonexistent, it can be assumed that the value of γ^* at which $\mu_L/\pi = 0.1$is the limit value of the velocity angle at which a solution of the problem is existent for a given wedge orientation.

Calculated entry angles γ^* versus the wedge rotation angle Ω are shown in Fig. 8 and compared with the experimental values of the velocity angle corresponding to the onset of flow separation at the wedge vertex reported in [30]. The figure also shows the entry angles at which the force coefficient on the left wedge side is zero (dashed line), which corresponds to the flow separation criterion proposed in [25].

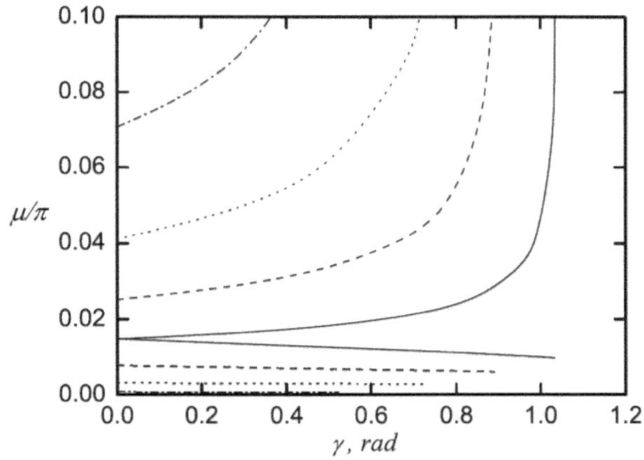

Fig. 7 Contact angles μ_R/π and μ_L/π on the right (decreasing curves) and the left (increasing curves) wedge side versus the entry velocity angle γ_∞ for wedge orientations $\Omega = 0$ (*solid lines*), $\Omega = 10°$ (*dashed lines*), $\Omega = 20°$(*dotted lines*), and $\Omega = 30°$ (*dash-dot lines*)

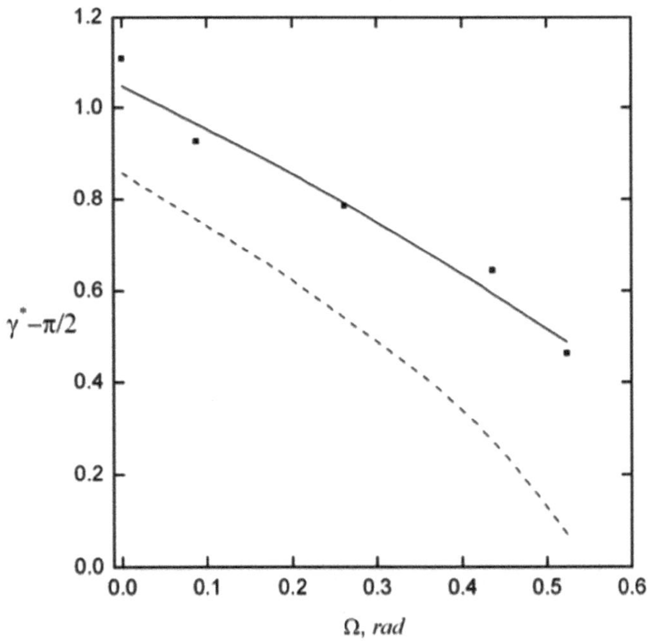

Fig. 8 Boundary of the region of nonseparated flow in the plane of the wedge rotation angle Ω and the wedge entry velocity angle γ^*: calculated results (*solid line*), experimental data [30] (*squares*), and the criterion $C_{nL}(\gamma^*, \Omega) = 0$ (*dashed line*). The region of nonseparated flow is below the lines

As illustrated, the criterion gives somewhat smaller wedge entry angles γ^* than those determined in the experiment.

Figure 9 shows the hydrodynamic force coefficients on the right (a) and the left (b) wedge sides versus the entry velocity angle for different wedge orientations with respect to the free surface.

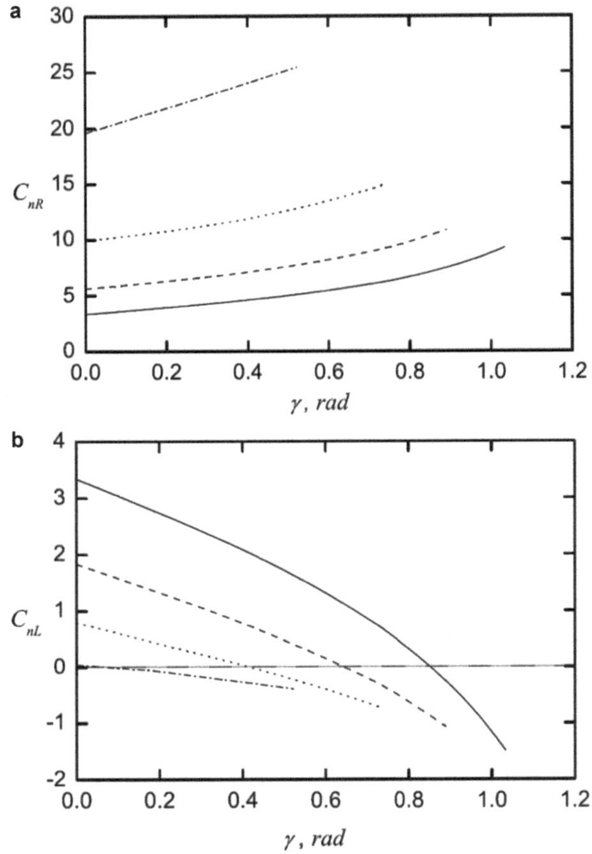

Fig. 9 Effect of the entry velocity angle γ_∞ on the normal force coefficients on the right (**a**) and the left (**b**) side of a $2\alpha = 106°$ wedge. Wedge rotation angle $\Omega = 0°$ (*solid lines*), $\Omega = 10°$ (*dashed lines*), $\Omega = 20°$ (*dotted lines*), and $\Omega = 30°$ (*dash-dot lines*).

For small values of γ, for which the velocity is nearly perpendicular to the unperturbed free boundary, the normal force coefficients vary nearly linearly: with increasing γ the normal force coefficient increases on the right wedge side (a) and decreases on the left one (b).

At some γ, the coefficient C_{nL} becomes negative. Negative values may occur if the average pressure $\bar{P}$ on the wedge side satisfies the condition $P_v < \bar{P} < P_a$ as can be seen from the expression $c_{pR} = (P - P_a)/\rho V_y^2$ whereP_v is the vapor pressure and V_y is the vertical component of the wedge entry velocity.

Conclusions

We have presented a generalized solution of the self-similar problem of impact between solid and liquid wedges, which includes as a special case the problem of the asymmetric and oblique entry of a wedge into a liquid. The solution method is based on constructing an analytical expression for the complex flow potential and finding a function that conformally maps the parameter region onto the flow region in the physical plane.

The flow geometry and the pressure distribution along the wedge sides are analyzed numerically over a wide range of given parameters. It is shown that at solid wedge angles greater than 180° the solution describes the cumulative effect, which consists in a many-fold increase in the liquid speed in the tip jets in comparison with the upstream speed.

The effect of the horizontal velocity component for wedge entry into a liquid with a flat free surface is studied, and the limit combinations of given parameters corresponding to flow separation from the wedge vertex are identified. The calculated and the experimental regions of nonseparated wedge entry in the velocity angle – wedge rotation angle plane are in good quantitative agreement.

For the limit combination of given parameters, the contact angle on the leeward side reaches its maximum value 18°, which is also typical for the symmetric entry of a wedge with a vertex angle approaching zero.

References

1. Korobkin AA, Pukhnachov VV. Initial stage of water impact. Ann Rev Fluid Mech. 1988;20:443–7.
2. von Karman T. The impact of seaplane floats during landing. Technical Note 321. Washington, DC: NACA; 1929.
3. Wagner H. Uber Stoss Gleitvorgange an der Oberache von Flussigkeiten. Z Angew Math Mech. 1932;12:192–215.
4. Garabedian PR. Oblique water entry of a wedge. Comm Pure Appl Math. 1953;6:157–65.
5. Borg SF. Some contributions to the wedge-water entry problem. Proc Am Soc Civil Engrs J Eng Mech Div. 1957;83(EM2):1214.
6. Logvinovich GV. Hydrodynamics of free-boundary flows (in Russian). Kiev: Naukova Dumka; 1969.
7. Moiseev NN, Borisova EP, Koryavov PP. Plane and axisymmetric self-similar problems of liquid entry (in Russian). Prikladnaya Matematika i Mekhanika. 1959;23:490–507.
8. Gonor AL. Entry of a thin wedge into a liquid (in Russian). Doklady Akademii Nauk SSSR. 1986;290:1068–72.
9. Mackie AG. The water entry problem. Q J Mech Appl Math. 1969;22:1–17.
10. Dobrovol'skaya ZN. Some problems of similarity flow of fluid with a free surface. J Fluid Mech. 1969;36:805–29.
11. Zhao R, Faltinsen O. Water entry of two-dimensional bodies. J Fluid Mech. 1993;246: 593–612.
12. Keady G, Fowkes N. The vertical entry of a wedge into water: integral equations and numerical results. Third Biennial Engineering Mathematics and Applications Conference (EMAC'98). Adelaide; 1998. Vol. 1, p. 277–81.
13. Freankel E. Problems of similarity flow of fluid with a free surface. J Fluid Mech. 1969;36:805–29.
14. Fraenkel LE, Keady G. On the entry of a wedge into water: the thin wedge and an all-purpose boundary-layer equation. J Eng Math. 2004;48(3):219–52.
15. Sagomonyan AYa. Penetration (in Russian). Moscow: Moscow State University; 1974.
16. Eroshin VA, Romanenkov NI, Serebryakov IV, Yakimov YuL. Hydrodynamic forces in blunt body entry into a compressible liquid (in Russian). Izvestiya Akademii Nauk SSSR, Mekhanika Zhidkostei i Gazov. 1980;6:44–51.
17. Kubenko VD. Entry of elastic shells into a compressible liquid (in Russian). Kiev: Naukova Dumka; 1981.

18. Korobkin AA. Blunt body entry into a weakly compressible liquid (in Russian). Zhurnal Prikladnoi Mekhaniki i Tekhnicheskoi Fiziki. 1984;5:104–10.
19. Zhuravlev YuF. Perturbation theory methods in spatial jet flows (in Russian). Trudy TsAGI. 1973;1532:1–22.
20. Savchenko YuN, Semenenko VN. Wave generation on the boundaries of cavities formed in the water entry of a disk and cones (in Russian). Problems in High-Speed Hydrodynamics. Cheboksary: Chuvash University; 1993. p. 231–9.
21. Savchenko YuN, Semenenko VN, Putilin SI. Unsteady processes in supercavitating body motion (in Russian). Prykladna Gidromekhanika. 1999;1(1):62–80.
22. Iafrati A. Hydrodynamics of asymmetric wedges impacting the free surface. European Congress on Computational Methods in Applied Sciences and Engineering. ECCOMAS 2000. Barselona; 11–14 Sept 2000.
23. Semenov YuA. Analytical solution to the self-similar problem of the water entry of an asymmetric wedge (in Russian). Prikladnaya Gidromekhanika. 2003;5(4, 77):64–72.
24. Semenov YuA. Oblique water entry of a wedge (in Ukrainian). Dopovidi Natsionalnoyi Akademiyi Nauk Ukrayiny. 2004;2:48–53.
25. Semenov YuA, Iafrati A. On the nonlinear water entry problem of asymmetric wedges. J Fluid Mech. 2006;547:231–56.
26. Goman OG, Semenov YA. Oblique entry of a wedge into an ideal incompressible fluid. Fluid Dynam. 2007;42(4):581–90.
27. Faltinsen OM, Semenov YA. Nonlinear problem of flat plate entry into an incompressible liquid. J Fluid Mech. 2008;611:151–73.
28. Chekin BS. Wedge entry into an incompressible liquid (in Russian). Prikladnaya Matematika i Mekhanika. 1989;53(3):396–404.
29. Korobkin AA. Inclined entry of a blunt profile into an ideal fluid. Fluid Dynam. 1988;23:443–7.
30. Judge C, Troesch AW, Perlin M. Initial water impact of a wedge at vertical and oblique angles. J Eng Math. 2004;48:279–98.
31. Xu L, Troesch AW, Vorus WS. Asymmetric vessel impact and planing hydrodynamics. J Ship Res. 1998;42:187–98.
32. de Divitiis N, de Socio LM. Impact of floats on water. J Fluid Mech. 2002;471:365–79.
33. Joukovski NE. Modification of Kirchhoff's method for determination of fluid motion in two directions at a fixed velocity given on the unknown streamline (in Russian). Math Coll. 1890; XV: 121–278.
34. Semenov YuA. Complex potential of an unsteady free-boundary flow (in Russian). Vestnik Khersonskogo Universiteta – Kherson. 2003;2:384–7.
35. Gurevich MI. Theory of Jets in ideal fluids (in Russian). Moscow: Nauka; 1979.
36. Wu GX. Two-dimensional liquid column and liquid droplet impact on a solid wedge. Q J Mech Appl Math. 2007;60:497–511.
37. Wu GX, Sun H, He YS. Numerical simulation and experimental study of water entry of a wedge in free fall motion. Roy Soc A, J Fluids Struct. 2004;19(3):277–89.
38. Semenov YA. Method for solving nonlinear problems on unsteady free-boundary flows.21th International Congress of Theoretical and Applied Mechanics. Warsaw; 15–21 August 2004.

Study of the Supercavitating Body Dynamics

V. N. Semenenko and Ye. I. Naumova

Abstract

In this paper, the results of investigations of dynamics of supercavitating (SC) bodies are presented, which were performed by authors in cooperation with Yu.N. Savchenko. Computer simulation of the SC-body motion based on the G.V. Logvinovich principle of independence of supercavity section expansion [1, 2] is the main research method. A general problem of the three-dimensional (3D) motion of the SC-body is formulated. Special cases of both the longitudinal and the lateral motion of SC-bodies are considered. Problems of the motion stability and optimization of SC-bodies moving on inertia on the arbitrary angle to the horizon are investigated. It is shown that the SC-vehicle motion in the regime of planing within a cavity is unstable on the depth. A comparative analysis of stabilization and control of motion (maneuverability) of SC-vehicles by inclination of the cavitator having two degrees of freedom and by the vectoring thrust is given.

In this paper, some materials from the works [3–9] were used, a part of the results were represented on the International conferences [10–13].

1 Introduction

As is known, dynamics of SC-bodies is essentially more complex than dynamics of bodies moving in the non-separated flow regime in air or in water. The complex unsteady behavior of a cavity formed by a body, and discrete interaction between the body and the cavity walls are causes of this complexity. The interaction force at each instant is defined by mutual position and relative motion of the body and the cavity, and the cavity shape is defined by the body motion prehistory (so called the memory effect of the unsteady cavity).

V.N. Semenenko (✉)
Institute of Hydromechanics of NASU, Kyiv, Ukraine
e-mail: Semenenko@ipnet.kiev.ua

I. Nesteruk (ed.), *Supercavitation*,
DOI 10.1007/978-3-642-23656-3_9, © Springer-Verlag Berlin Heidelberg 2012

The corresponding problems are strictly non-linear, they contain discontinuous functions and functions with a lagging argument. The numerical analysis of such complex dynamic systems, which is based on a complete set of equations, is the unique reliable method of them investigation.

For the computer simulation of the SC-body motion we use the mathematical model based on the fundamental results of the SC-flow investigations obtained in USSR in 1960–1980s. They were collected and generalized in lectures by Yu.N. Savchenko and V.N. Semenenko [14–18], which were given in February 2001 at the von Carman Institute for Fluid Dynamics in Brussels, Belgium. As the previous 10 years have shown, these lectures influenced positively upon recognition and distribution of the G.V. Logvinovich's theory and approximation methods of calculation of the cavity shape [1, 2] and also forces when a body is planing within the cavity [19] among the foreign specialists. At present, this mathematical model is used practically in all the works on the SC-vehicle control. The basic principles and methods of control of the SC-vehicle motion were stated in Yu.N. Savchenko's papers [15, 20]. Lately, many publications devoted to the SC-body dynamics and control appeared. A brief review of papers [21–34] published after 2000 are given below.

The results of calculating the SC-model motion on inertia, which were obtained on a base of G.V. Logvinovich's simplified equations, are given in paper [21]. We make the remark that the oscillatory behavior of the model motion obtained in [21] has been experimentally discovered by us earlier [3] and confirmed by the computer simulation [4, 5]. In papers [22, 23], the attempts of calculation of 3D motion of SC-bodies by the CFD methods were made. As the practice showed, these methods are still too labor-consuming and ineffective as applied to the SC-body dynamic investigation. Therefore, the approximate model of the SC-body motion based on the G.V. Logvinovich's theory [1, 19] is used in papers [24, 25] and in the all the next papers.

As is known, the SC-vehicle motion in the regime of planing within the cavity is unstable on the depth [13]. The method of active stabilization of such motion by automatic regulating the disk cavitator inclination at the linear feedback law is discussed in papers [25, 26]. The more complex problem of constructing the non-linear automatic control system for the SC-vehicle is formulated in paper [26] as well. In further this approach was developed by a number of other researchers.

Both the depth maneuvering and the course maneuvering of the SC-vehicle as particular cases of 3D motion are considered in papers [25, 27, 28]. The rotary cavitator and traditional fins and rudders working as SC-hydrofoils are used as the operating control. The paper [28] proposes the method of approximate taking into consideration the cavity memory effect by means of introducing the lagging argument in the equation of the SC-vehicle dynamics. A number of other papers of the same authors are devoted to the practically important problems of the SC-body strength and their construction optimization.

The papers [29, 30] describe various methods of synthesis of the automatic control systems intended to optimize the SC-vehicle trajectory. The SC-vehicle behavior in motion in the vertical plane from the point of view of the nonlinear theory of dynamic systems (instability, bifurcation, chaos) is considered in papers [31, 32]. A possibility of application of different more complex classes of the robust, relay, and adaptive automatic control systems for the depth stabilization of the SC-vehicles is discussed in papers [32–34]. One notes that in addition to [33, 34] a number of papers of the Chinese authors considering analogous problems were published lately.

The most of the mentioned papers about the SC-body dynamics and the SC-vehicle control have the following common features:

1. The simplified mathematical model of the SC-vehicle is used, which does not take into account the unsteady cavity memory effect
2. The longitudinal motion of the SC-vehicle i.e. motion in the vertical plane is considered (except [25, 27, 28])
3. The motion control by inclination of the disk cavitator having one degree of freedom is considered
4. The synthesis of the automatic control systems is realized by the traditional method of linearization of the dynamic equations

The main distinctions of this paper from the mentioned above ones consist in the following:

1. The more adequate mathematical model is used, that naturally takes into account the unsteady cavity memory effect
2. Both the longitudinal motion and the lateral motion of a SC-vehicle is considered
3. The control of the SC-vehicle motion by inclination and/or turning the cavitator having the two degrees of freedom (δ-control) is considered
4. The control of the SC-vehicle motion by the vectoring thrust (η-control) is considered
5. The action of the automatic control systems is modeled basing on the complete nonlinear equations of the SC-vehicle dynamics

2 Equations of 3D Motion of SC-Body

Figure 1 shows a scheme of the 3D SC-body motion. The body coordinate system $O_1x_1y_1z_1$ and the flow coordinate system $O_1x_0y_0z_0$ are shown. An origin of both the coordinate systems is placed at the vehicle mass center O_1. The axis O_1x_1 of the body coordinate system is directed along the longitudinal vehicle axis. The axis O_1x_0 of the flow coordinate system is directed along the velocity vector $\vec{V}$ of the vehicle mass center. The axes O_1y_1 and O_1y_0 are placed in the diametral plane of the body.

Also, we will use the fixed coordinate system $Oxyz$ and the semi-body coordinate system $O_1x_gy_gz_g$. The direction of the semi-body coordinate system axes coincides with direction of axes of the fixed coordinate system $Oxyz$ at each time instant.

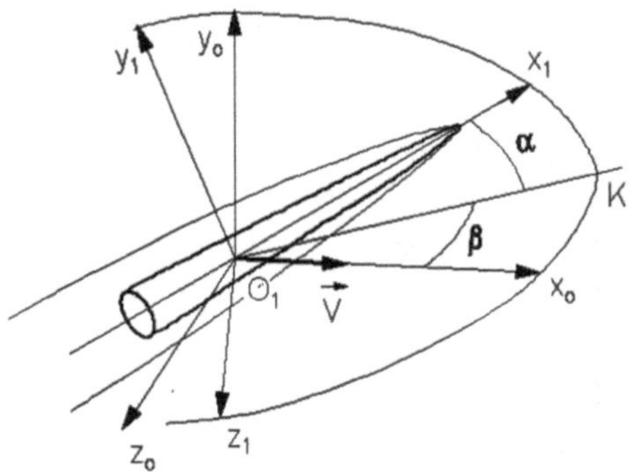

Fig. 1 Scheme of 3D motion of SC-body and the coordinate systems

One writes a set of equations of the 3D motion of a solid body in projections on the axes of the body coordinate system $O_1x_1y_1z_1$, which are the principal axis of inertia of the body [35]:

$$m\left(\frac{dV_{x1}}{dt} + \omega_y V_{z1} - \omega_z V_{y1}\right) = F_{x1}, \tag{1}$$

$$m\left(\frac{dV_{y1}}{dt} + \omega_z V_{x1} - \omega_x V_{z1}\right) = F_{y1}, \tag{2}$$

$$m\left(\frac{dV_{z1}}{dt} + \omega_x V_{y1} - \omega_y V_{x1}\right) = F_{z1}, \tag{3}$$

$$I_x\frac{d\omega_x}{dt} + \omega_y\omega_z(I_z - I_y) = M_{x1}, \tag{4}$$

$$I_y\frac{d\omega_y}{dt} + \omega_x\omega_z(I_x - I_z) = M_{y1}, \tag{5}$$

$$I_z\frac{d\omega_z}{dt} + \omega_x\omega_y(I_y - I_x) = M_{z1}, \tag{6}$$

$$\omega_x = \frac{d\theta}{dt} + \frac{d\varphi}{dt}\sin\psi, \tag{7}$$

$$\omega_y = \frac{d\varphi}{dt}\cos\psi\cos\theta + \frac{d\psi}{dt}\sin\theta, \tag{8}$$

$$\omega_z = \frac{d\psi}{dt}\cos\theta - \frac{d\varphi}{dt}\cos\psi\sin\theta, \tag{9}$$

where m is the body mass; $\vec{V} = \{V_{x1}, V_{y1}, V_{z1}\}$ is the velocity vector of the body mass center; $\vec{\omega} = \{\omega_x, \omega_y, \omega_z\}$ is the angular velocity vector relatively to the body mass center; θ is the roll angle; φ is the yaw angle; ψ is the pitch angle; I_x, I_y, I_z are the moments of inertia relatively to the axes O_1x_1, O_1y_1, O_1z_1, respectively; F_{x1}, $F_{y1}, F_{z1}, M_{x1}, M_{y1}, M_{z1}$ are the projections of the resultant force vector and the main moment on the same axes.

The body mass center trajectory relatively to the fixed coordinate system $Oxyz$ is defined by equations:

$$\frac{dx}{dt} = V\cos(\psi - \alpha)\cos(\varphi - \beta), \tag{10}$$

$$\frac{dy}{dt} = V\sin(\psi - \alpha)\cos(\varphi - \beta), \tag{11}$$

$$\frac{dz}{dt} = -V\sin(\varphi - \beta), \tag{12}$$

where $V = |\vec{V}|$; α is the angle of attack; β is the sliding angle. The angles α and β define a position of the body coordinate system relatively to the flow coordinate system (see Fig. 1). In this case the following relations are valid [36]:

$$V_{x1} = V\cos\alpha\cos\beta, \quad V_{y1} = -V\sin\alpha\cos\beta, \quad V_{z1} = V\sin\beta. \tag{13}$$

One accepts the following assumptions for formulation of the general problem of the SC-body dynamics:

1. The SC-body is a slender body of revolution, in this case $I_y = I_z$
2. A disk with diameter D_n is the cavitator shape
3. The mass m, the mass center position x_c, and the vehicle moments of inertia I_x, I_z do not vary during motion

Then the right parts of Eqs. 1–3 must include the projections of the gravity force mg, the hydrodynamic force on the cavitator $\vec{F}_n$, the force of interaction between the body and the cavity wall $\vec{F}_s$, and also the propulsor thrust $\vec{F}_{pr}$ and the control force on the fins and rudders $\vec{F}_c$, if they are present:

$$\begin{aligned} F_{x1} &= F_{nx} + F_{sx} - mg\sin\psi + F_{pr}\cos\eta_z\cos\eta_y + F_{cx}, \\ F_{y1} &= F_{ny} + F_{sy} - mg\cos\psi\cos\theta + F_{pr}\sin\eta_z\cos\eta_y + F_{cy}, \\ F_{z1} &= F_{nz} + F_{sz} + mg\cos\psi\sin\theta - F_{pr}\sin\eta_y + F_{cz}, \end{aligned}$$

where $F_{pr} = |\vec{F}_{pr}|$; η_y and η_z are the angles of deflection of the thrust vector in the planes $O_1x_1z_1$ and $O_1x_1y_1$, respectively. The right parts of Eqs. 4 and 5 must include the projections of moments of the corresponding forces relatively to the body mass center. Figure 2 shows a scheme of the forces acting onto the SC-vehicle without

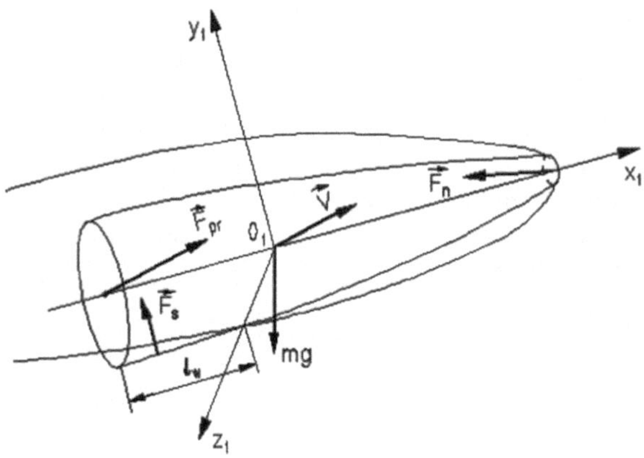

Fig. 2 Scheme of forces acting onto SC-vehicle in the regime of planing within the cavity

fins moving under action of the propulsor thrust in the regime of planing within the cavity.

The set of the differential equations (1)–(12) is integrated numerically at the initial conditions:

$$\begin{aligned}
&V_{x1}(0) = V_0 \cos\psi_0 \cos\varphi_0,\ V_{y1}(0) = -V_0 \sin\psi_0 \cos\varphi_0,\\
&V_{z1}(0) = V_0 \sin\varphi_0,\ \omega_x(0) = \omega_{x0},\ \omega_y(0) = \omega_{y0},\\
&\omega_z(0) = \omega_{z0},\ \theta(0) = \theta_0,\ \varphi(0) = \varphi_0,\\
&\psi(0) = \psi_0,\ x(0) = 0,\ y(0) = 0,\ z(0) = 0.
\end{aligned}$$

3 Calculation of Unsteady Cavity

The equation of expansion of the axisymmetric cavity sections [17, 37] is used to calculate the unsteady cavity shape. It is expression of the G.V. Logvinovich's principle of independence of the cavity section expansion [1]:

$$\frac{\partial^2 S_c(\tau,\ t)}{\partial t^2} = -\frac{k_1 \Delta p(\tau)}{\rho},\quad s(t) - L_c(t) \le \xi \le s(t)\,, \tag{14}$$

where S_c is the cavity section area (see Fig. 3); s is the arc coordinate of the cavitator center; k_1 is the semi-empirical constant; $\tau \le t$ is the time instant of the cavitator passage through the section ξ; L_c is the cavity length; p_c is the cavity pressure; $\Delta p(\tau) = p_\infty(\xi) - p_c$. According to the independence principle the cavity axis shape is defined by the trajectory of the cavitator center (x_n, y_n, z_n):

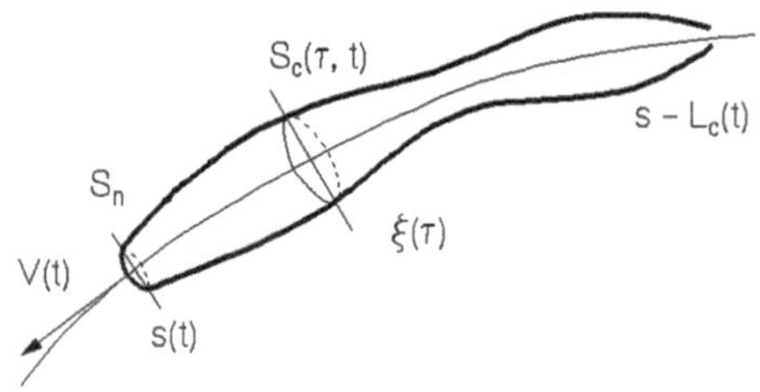

Fig. 3 The calculation scheme of unsteady supercavity

$$x_n(t) = x(t) + x_c \cos\varphi(t) \cos\psi(t),$$
$$y_n(t) = y(t) + x_c \sin\psi(t),$$
$$z_n(t) = z(t) - x_c \sin\varphi(t),$$

where (x, y, z) are the current coordinates of the body mass center; x_c is the distance from the cavitator to the mass center.

The cavity axis may be additionally curved under action of the lateral force arising on the inclined cavitator and under action of the gravity forces as well. For the steady cavities these effects are approximately taken into account by the approximate formulae [38]:

$$h_{fy}(x) = c_{ny0} R_n \,(0.46 - \sigma + \bar{x}/2),$$
$$h_{fz}(x) = -c_{nz0} R_n \,(0.46 - \sigma + \bar{x}/2),$$
$$h_{gy}(x) = \frac{(1+\sigma)\,x^2}{3L_c\,Fr_l^2}, \quad Fr_l = \frac{V}{\sqrt{gL_c}},$$

where R_n is the cavitator radius; c_{ny0} and c_{nz0} are the coefficients of the forces F_{ny0} and F_{nz0}; $\bar{x} = x/L_c$. As it was established [38], the cavity axis curving is the main perturbations of the cavity shape, and deformations of the circular shape of the transversal cavity sections may be neglected if Fr is large. Thus, owing to the independence principle at each time instant the coordinates of contours of the axial cavity section in the fixed coordinate system are determined by relations

$$y_{cav}(\tau, t) = \pm R_c(\tau, t) + y_n(\tau) + h_{fy}(\tau, t) + h_{gy}(\tau, t),$$
$$z_{cav}(\tau, t) = \pm R_c(\tau, t) + z_n(\tau) + h_{fz}(\tau, t), \tag{15}$$

where $R_c = \sqrt{S_c/\pi}$ is the radius of the current cavity section.

The mathematical model (14), (15) takes into consideration the unsteady cavity memory effect in the natural way. We repeatedly checked and specified it by comparison and agreement with the experimental results. The experiments were performed for "small" and "big" models in the wide range of the flow velocities on the high-speed multi-purpose hydrodynamic tunnel at the Institute of Hydromechanics of NAS of Ukraine (see for example [3, 14, 17]).

4 Determination of Forces and Moments Acting Onto SC-Body

The cavitator is the most important element from the point of view of the SC-body dynamics, since it serves not only to forming a cavity, but to control of the motion. In addition to the force on the cavitator $\vec{F}_n$, the force of interaction between the body and the cavity walls $\vec{F}_s$ is of importance. The gravity mg, the propulsor thrust $\vec{F}_{pr}$, and the force $\vec{F}_c$ created by fins and rudders (if they are present) are concerned to other forces acting onto the SC-body.

4.1 The Force on the Inclined Cavitator

Let the cavitator is oriented in arbitrary way in relation to the coordinate system $O_1x_1y_1z_1$. One introduces the cavitator coordinate system $O_nx_ny_nz_n$ (see Fig. 4). The origin of coordinates O_n is placed on the longitudinal body axis O_1x_1; the axis O_nx_n is directed along the normal $\vec{n}$ to the cavitator plane in direction to the liquid; the axis O_ny_n is placed in the plane $O_1x_1y_1$; δ_y is the angle between the axis O_nx_1 and the projection O_nK of the axis O_nx_n onto the plane $O_nx_1z_1'$ (where $O_nz_1' || O_1z_1$); δ_z is the angle between the axis O_nx_n and the plane $O_nx_1z_1'$.

In the case of the disk cavitator the force from the flow side is always directed oppositely to the normal $\vec{n}$ to the cavitator plane: $\vec{F}_n = -F_n\vec{n}$. If the normal to the cavitator is inclined to the flow on the angle μ, then the absolute value of the force acting to the cavitator is equal to [1]

$$F_n = \left|\vec{F}_n\right| = F_{x0}\cos\mu, \tag{16}$$

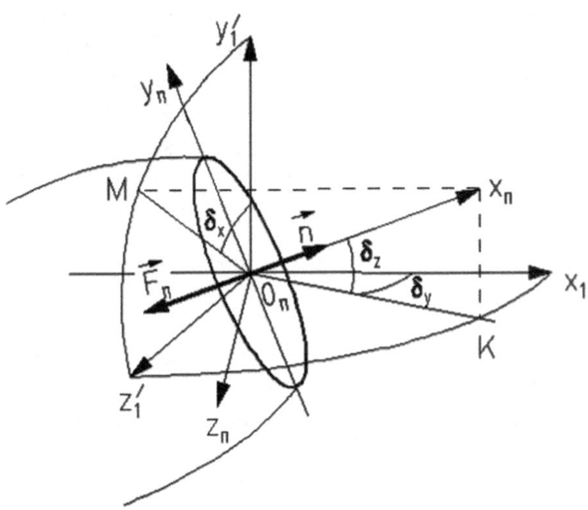

Fig. 4 The coordinate system of inclined cavitator

where $F_{x0} = 0,82\ (1+\sigma)$ is the cavitation drag of the disk oriented perpendicularly to the flow. In our paper [9] the following expression was obtained:

$$\cos\mu = c_{11}^{(n)} = \cos\delta_y \cos\delta_z \cos\alpha_n \cos\beta_n - \\ -\sin\delta_z \sin\alpha_n \cos\beta_n - \sin\delta_y \cos\delta_z \sin\beta_n, \tag{17}$$

where $\alpha_n = \alpha + \omega_z x_c / V_{x1}$, $\beta_n = \beta + \omega_y x_c / V_{x1}$. The projections of the vector $\vec{F}_n$ on the axes of the body coordinate system are equal to:

$$F_{nx} = -F_n \cos\delta_y \cos\delta_z, \quad F_{ny} = -F_n \sin\delta_z, \quad F_{nz} = F_n \sin\delta_y \cos\delta_z. \tag{18}$$

The projections of the same vector on the axes of the flow coordinate system are equal to [9]:

$$F_{nx0} = -F_n c_{11}, \quad F_{ny0} = -F_n c_{12}, \quad F_{nz0} = -F_n c_{13}, \tag{19}$$

where

$$\begin{aligned} c_{11} &= \cos\delta_y \cos\delta_z \cos\alpha \cos\beta - \\ &\quad - \sin\delta_z \sin\alpha \cos\beta - \sin\delta_y \cos\delta_z \sin\beta, \\ c_{12} &= \cos\delta_y \cos\delta_z \sin\alpha + \sin\delta_z \cos\alpha, \\ c_{13} &= -\cos\delta_y \cos\delta_z \cos\alpha \sin\beta + \\ &\quad + \sin\delta_z \sin\alpha \sin\beta - \sin\delta_y \cos\delta_z \cos\beta. \end{aligned}$$

Any position of the cavitator may be obtained by realizing the two consequential turnings of the cavitator on the angle δ_z' around the axis $O_n z_1'$ and on the angle δ_x around the axis $O_1 x_1$. In this case a connection between the pairs of angles (δ_z', δ_x) and (δ_y, δ_z) is given by the relations

$$\cos\delta_z' = \cos\delta_y \cos\delta_z, \quad \sin\delta_x = \frac{\sin\delta_y \cos\delta_z}{\sin\delta_z'}. \tag{20}$$

The first method of setting the cavitator orientation is convenient when solving the dynamic equations, the second method is the more convenient when practical realization of the SC-vehicle control system.

For the axisymmetric SC-vehicle without fins, the turning the cavitator on the angle δ_x about the axis $O_1 x_1$ is equivalent to rolling of the whole vehicle on the angle $\theta = \delta_x$. Thus, for such a vehicle the δ-control on course is equivalent to the "bank-to-turn" control [25]. In this case recalculation of the angles can be performed by formulae (20).

4.2 The Planing Force

The force F_{sy} of planing of an elongated body within a cavity may be calculated by the Wagner's method [1] with using the solution of the two-dimensional problem on immersion of a circular arc in a curvilinear free surface [19, 39]. In the case of motion of the SC-body in the vertical plane it can be calculated at each time instant by the formula:

$$F_{sz} = \rho \pi R_s^2 V_0 \left[V_1 \frac{\bar{h}(2+\bar{h})}{(1+\bar{h})^2} + V_2 \frac{2\bar{h}}{1+\bar{h}} \right], \tag{21}$$

where $V_1 = \alpha_s V - V_y + \omega_z (L - x_c) + V_{yc}$ is the vertical velocity of the body transom; L is the body length; V_y is the vertical velocity of the mass center; V_{yc} is the transversal velocity of the cavity axis; V_2 is the velocity of relative motion of the cavity wall and body; $\bar{h} = h/(R_c - R_s)$; h is the immersion of the body tail edge; R_c, R_s are the radii of the cavity and body at the tail edge, respectively.

The longitudinal component of the planing force F_{sx} has viscous nature, it is calculated by the formula:

$$F_{sx} = \frac{\rho V^2}{2} S_w c_f(\mathrm{Re}_w), \quad \mathrm{Re} = \frac{l_w V}{\nu}, \tag{22}$$

where c_f is the friction drag coefficient [40]; l_w and S_w are the length and the area of the wetted body surface, respectively.

Formulae (21), (22) may be applied in the case of arbitrary 3D motion of the body within the cavity. In this case one should consider that the force $\vec{F}_s$ is applied at each time instant in that point of the contact arc where the immersion h is maximal, and it is directed so that the force vector passes through the body section center.

4.3 Other Forces and Moments

If the mass center coincides with the body transversal section center, then the gravity mg does not create any moment. If the mass center is placed below the section center on the distance h_M (so called the metacentric height), then when rolling the vehicle on the angle θ the restoring moment relatively to the axis O_1x_1 arises:

$$M_{gx} = -h_M mg \cos\psi \sin\theta. \tag{23}$$

Using the traditional operating controls piercing the cavity walls (fins and rudders) for SC-vehicles is ineffective by a number of reasons [10]. In this paper, it is considered that the fins and rudders are absent, i.e. $\vec{F}_c = 0$. However, it is obvious that the effect of control of the SC-vehicle motion by the tail rudders will be qualitatively similar to the effect of control by deflecting the propulsor thrust vector $\vec{F}_{pr}$ considered below.

5 Two Classes of SC-Models

It is convenient to sort the whole multitude of the investigated SC-bodies on two classes: (1) the "small" high-speed SC-models moving in water on inertia on the arbitrary angle to the horizon and (2) the "big" SC-vehicles moving under the propulsor thrust. Owing to distinctions in velocities and motion conditions the optimal shape and dynamic behavior of the models of the first and second classes can be essentially different.

Absence of both the propulsor and the active control, i.e. $\vec{F}_{pr} = \delta_y = \delta_z = 0$ is typical for models of first class. The problems of the motion stability and achievement of the maximal range are the most important problems for models of first class. The problems of the motion stabilization and controllability (maneuverability) are the most important problems for models of second class.

5.1 The Calculation Model of the "Small" SC-Body

Since the range of the SC-body motion on inertia is proportional to the model mass m, then the shape of the "small" SC-model at the given density ρ_b must be inscribed maximally tightly into the frontal cavity part. Besides, the model shape must ensure the motion stability.

Examples of calculations for the uniform model are given below. Its shape and dimensions are shown in Fig. 5 and it has the following parameters: $m = 107$ g, $\bar{x}_c = x_c/L = 0.67$, $I_z = 13.27 \cdot 10^{-5}$ kg m^2.

5.2 The Calculation Model of the SC-Vehicle

Figure 6 shows the shape and dimensions of the SC-vehicle model designed in such way that on the marching part of the flying path the model moves in the regime of planing within the cavity. The model mass is $m = 600$ g, the moments of inertia are $I_x = 8$ kg·m^2, $I_y = I_z = 900$ kg·m^2. It is supposed that the model moves

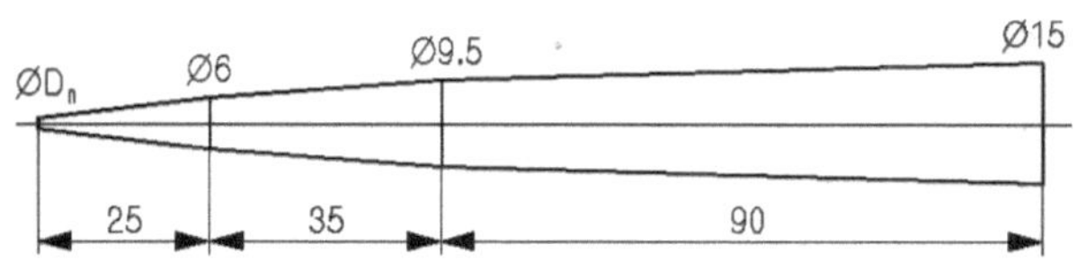

Fig. 5 The calculation scheme of "small" SC-model

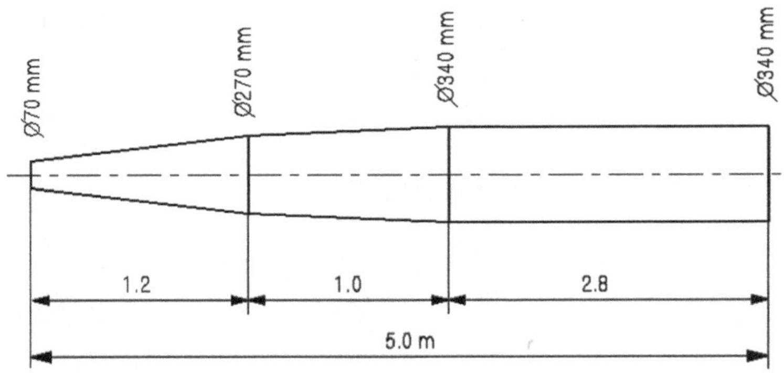

Fig. 6 The calculation scheme of SC-vehicle

horizontally on the marching part of the flying path on the depth $H_m = 5\text{m}$ with the constant velocity $V_m = 120$ m/s. In this case the cavitation number and the Froude number are equal to $\sigma = 0.0201$ and $Fr = 144.8$, and the cavity length $L_{cm} = 6.52$ m. For the accepted parameters the calculation gives the value of the required propulsor thrust equal to the total drag of the SC-vehicle $F_m = 23.18$ KN.

6 Equations of Longitudinal Motion of SC-Body

As in the case of aircrafts and traditional underwater vehicles, the general 3D motion of the SC-body may be approximately divided on the longitudinal motion and the lateral motion [36]. The longitudinal motion is the motion in the vertical plane (revolution around the axis O_1z_1 and transition in direction to the axes O_1x_1 and O_1y_1). It is described by a set of four Eqs. 1, 2, 6, and 9, in which it is necessary to accept $\theta = 0$, $\varphi = 0$, $\omega_x = 0$, $\omega_y = 0$. The SC-body longitudinal motion may be investigated independently on the lateral motion.

At formulation of the problem on the longitudinal motion we accept the following assumptions additionally to the mentioned ones:

1. The motion of the SC-body occurs in the vertical plane (i.e. $V_{z1} = 0$, $\omega_y = 0$, $\varphi = 0$, $\beta = 0$, $z = 0$)
2. The body revolution about the longitudinal axis is absent (i.e. $\omega_x = 0$, $\theta = 0$)

Then, six Eqs. 1, 2, 6, 9–11 remain from the general set of dynamic equations (1)–(12). Passing to differentiating with respect to the absolute coordinate x, one obtains the calculation scheme of the five differential equations of the SC-body longitudinal motion:

$$V\cos(\psi - \alpha)\frac{dV_{x1}}{dx} = \omega_z V_{y1} + \frac{1}{m}F_{x1}, \tag{24}$$

$$V\cos(\psi - \alpha)\frac{dV_{y1}}{dx} = -\omega_z V_{x1} + \frac{1}{m}F_{y1}, \tag{25}$$

$$V\cos(\psi-\alpha)\frac{d\omega_z}{dx}=\frac{1}{I_c}M_{z1}, \tag{26}$$

$$V\cos(\psi-\alpha)\frac{d\psi}{dx}=\omega_z, \tag{27}$$

$$\frac{dy}{dx}=\mathrm{tg}(\psi-\alpha), \tag{28}$$

where $V=\sqrt{V_{x1}^2+V_{y1}^2}$; $\alpha=-\mathrm{arctg}\frac{V_{y1}}{V_{x1}}$. For any x the elapsed time can be calculated by the formula:

$$t(x)=\int_0^x \frac{ds}{V\cos(\psi-\alpha)}.$$

In the case of the longitudinal motion, the right parts of Eqs. 24–26 and relations (18), (19) take the following form:

$$\begin{aligned}F_{x1}&=F_{nx}+F_{sx}-mg\sin\psi+F_{pr}\cos\eta_z,\\ F_{y1}&=F_{ny}+F_{sy}-mg\cos\psi+F_{pr}\sin\eta_z,\\ M_{z1}&=M_n+M_s+M_{pr}.\end{aligned}$$

$$\begin{aligned}F_{nx}&=-F_{x0}\cos(\alpha_n+\delta_z)\cos\delta_z,\\ F_{ny}&=-F_{x0}\cos(\alpha_n+\delta_z)\sin\delta_z.\end{aligned} \tag{29}$$

$$\begin{aligned}F_{nx0}&=-F_{x0}\cos(\alpha_n+\delta_z)\cos(\alpha+\delta_z),\\ F_{ny0}&=-F_{x0}(\alpha_n+\delta_z)\sin(\alpha+\delta_z).\end{aligned} \tag{30}$$

The set of differential equations (24)–(28) is integrated numerically at the initial conditions:

$$\begin{aligned}&V_{x1}(0)=V_0\cos\psi_0,\quad V_{y1}(0)=-V_0\sin\psi_0,\\ &\omega_z(0)=\omega_{z0},\quad \psi(0)=\psi_0,\quad y(0)=0.\end{aligned}$$

7 The Longitudinal Motion of SC-Bodies Moving on Inertia

Let the body starts on the depth H_0 with velocity V_0 on an arbitrary angle to the horizon γ ($\gamma<0$ at downstream motion). In this case it is convenient to define the fixed coordinate system Oxy by directing the axis Ox along the starting velocity vector. Then the hydrostatic water pressure in formula (14) will be changed according to the formula:

$$p_\infty(\tau) = p_{atm} + \rho g[H_0 - x(\tau)\sin\gamma - y(\tau)\cos\gamma],$$

where $y(\tau)$ is the deflection of the cavity section center formed at the time instant τ from the axis Ox.

7.1 Stability of the "Small" SC-Model Motion

The computer simulation has confirmed that the self-stabilization of "small" SC-models by ricocheting the model tail from the cavity walls is the basic mechanism of the motion stability [3–5]. In this case the degree of the SC-model motion stability considerably depends on the model shape and its mass center position.

Figure 7 shows graphs of dependencies of the motion range S on the depth H calculated for three cavitator diameters when $\gamma = 0$; $V_0 = 1000$ m/s and $\omega_{z0} = 2.0$ rad./s. As is obvious, for the low H independently on the value D_n the SC-model passes a distance about 10 m and then loses the stability. With increasing H the motion becomes stable, and the range spasmodically increases.

The zones of stable and unstable motions on the plane of parameters (H, ω_{z0}) for various D_n are shown in Fig. 8. For each of three cavitator diameters D_n the curve divides the plane on two zones. The stable motion zone is placed to the right from the curve, The unstable motion zone is placed to the left.

A comparison of the graphs in Figs. 7 and 8 shows that the SC-model motion is more stable for smaller D_n and higher H, i.e. with the tighter cavity. The analogous conclusion was made earlier in paper [41].

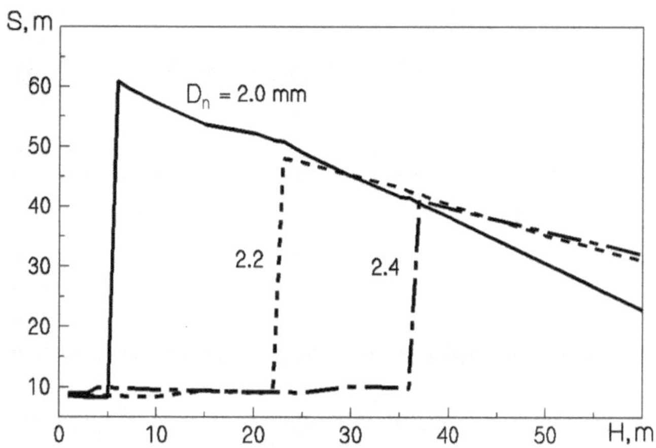

Fig. 7 Range versus the motion depth

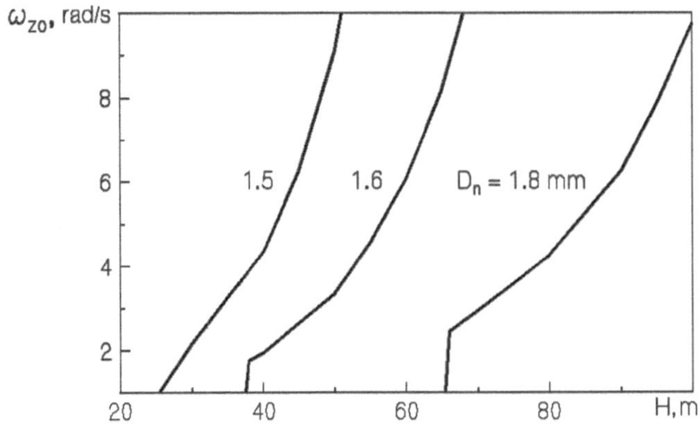

Fig. 8 Zones of the stable and unstable motions of SC-model

7.2 The Optimization Problems for SC-Bodies Moving on Inertia

During inertial motion of SC-bodies their velocity rapidly decreases, and the cavity dimensions decrease too. As the experience shows, the motion continues until the body is fully covered with the cavity and practically instantly ceases when the body is wetted by water [3]. Many parameters influences on the motion range: the body shape, the body length, the body mass, the cavitator diameter, the starting velocity, the depth and the angle of launching etc. Naturally, the optimization problem arises: how to choose the body shape and other parameters that the range of the SC-body motion on inertia would be maximal. Putting aside the motion stability problem here, one considers two typical optimization problems for SC-bodies moving rectilinearly.

Task 1. To choose the cavitator diameter D_n for the body with the given shape and mass so that the range S would be maximal. Figure 9 shows the calculation results when $H_0 = 20$ m; $V_0 = 500$ m/s, for a number of values of the starting angle γ. The calculation shows that the optimal values of D_n (marked by circles) depend on the angle γ and on the motion depth and do not depend on the model mass.

Task 2. To choose the optimal ratio of the mass m and the starting velocity V_0 for given both the cavitator diameter D_n and the model initial kinetic energy $E_0 = mV_0^2/2$ so that the range S would be maximal. Figure 10 shows the calculation results when $D_n = 2.2$ mm; $E_0 = 13.392$ kJ; $\gamma = 0$. As one can see, the optimal ratio of the model mass and the starting velocity when $E_0 =$ const also depends on the motion depth.

In the case of low cavitaton numbers and high Froude numbers we developed the approximate analytical method of optimization of parameters of the SC-bodies moving on inertia [6, 7]. The obtained simple relations allow the ultimately possible

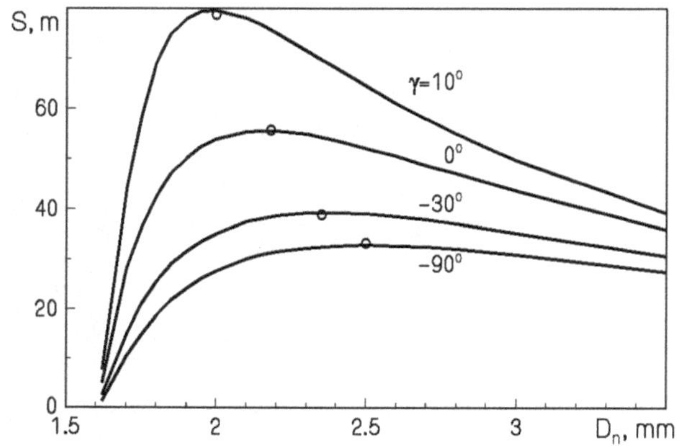

Fig. 9 Range versus the cavitator diameter for various γ

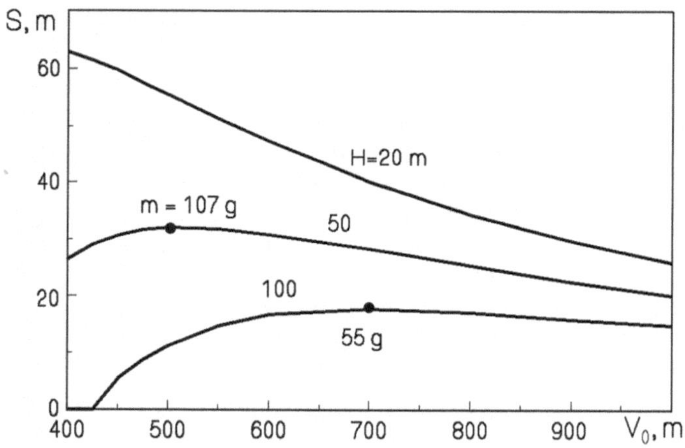

Fig. 10 Range versus the starting velocity when E_0= const

range of the SC-body motion in water to be estimated. A number of specific problems of optimization of the SC-bodies moving on inertia on the arbitrary angle to the horizon were solved by this method in papers [6, 7, 42] for wide variation of isoperimetric conditions (also see I. Nesteruk's article in this book).

We have shown that the range of the SC-body motion on inertia may be increased by means of dynamic changing its parameters during the motion. The transformer cavitators having the variable drag [3] or the variable diameter may be applied for this purpose. So, properly increasing the cavitator diameter during motion, one can obtain the range increasing up to 60% [7].

8 Longitudinal Motion of SC-Vehicles

In this case we define the fixed coordinate system Oxy directing the axis Ox horizontally. One considers some problems arising at practical application of self-propelled SC-vehicles and successfully solvable in the frames of model of the longitudinal motion.

8.1 The SC-Vehicle Acceleration

One of the main problems consists in that the vehicle must overcome the very high friction drag during its acceleration up to formation of a full-developed supercavity. On the acceleration part the vehicle moves sequentially in the non-separated regime, in the partial cavition regime and, finally, in the regime of planning within the cavity. The different mechanisms of forming the motion drag act on each of these three parts.

Figure 11 shows an example of calculation of the cavity length increasing during the vehicle acceleration under action of the fixed propulsor thrust F_m. The graphs of increasing the vehicle velocity look analogously [13]. Curve 1 corresponds to the thrust value $F_{pr} = F_m$. As is obvious, the marching values of the cavity length and the velocity (in this case $L_{cm} = 8.71$ m, $V_m = 150$ m/s) are not achieved, and the attained balanced values $L_c = 4.93$ m, $V = 112.5$ m/s correspond to the partial cavitation regime. When the thrust increases (curve 2) the vehicle reaches the balanced values L_c and V exceeding considerably the marching values L_{cm} and V_m.

A dependence of the final balanced cavity length after acceleration on the relative propulsor thrust F_{pr}/F_m is shown in Fig. 12. The graph of dependence of the final velocity of the vehicle on F_{pr}/F_m looks analogously. As is obvious, transition from the partial cavitation regime to the supercavitation with increasing

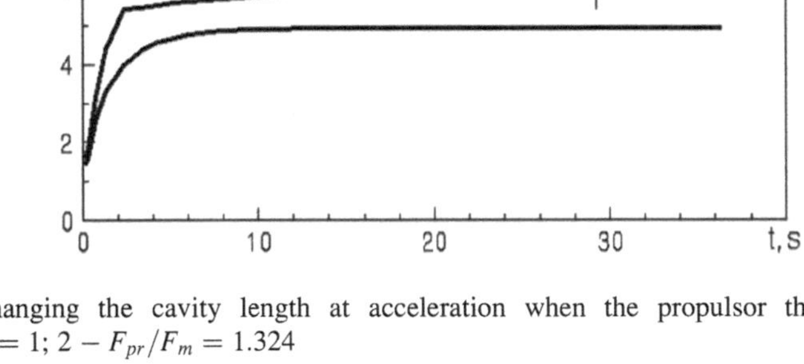

Fig. 11 Changing the cavity length at acceleration when the propulsor thrust is fixed $1 - F_{pr}/F_m = 1$; $2 - F_{pr}/F_m = 1.324$

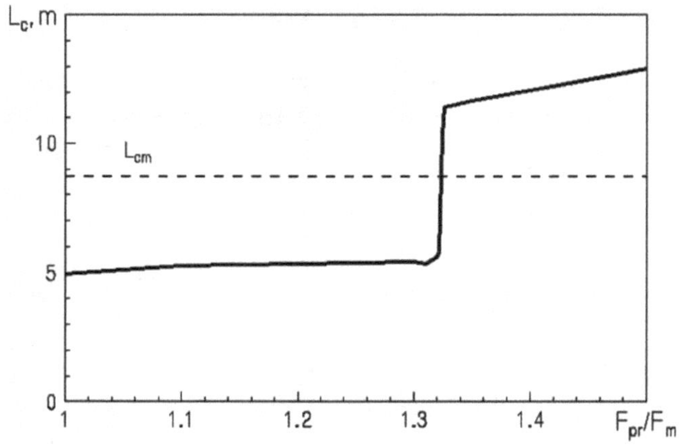

Fig. 12 The final cavity length versus the relative thrust

the thrust occurs stepwise. In this case the vehicle attains uneconomical flow regimes when $L_c > L_{cm}$, $V > V_m$.

Thus, it is impossible to come to the marching motion regime of the SC-vehicle using only the mid-flight engine with the fixed thrust F_m beginning from the low starting velocity. The two methods of solving this problem exist:

1. Using the additional starting propulsor (accelerator) generating very high thrust $F_{st} >> F_m$ during short time
2. Using the intensive gas-supply into a partial cavity with the purpose of its development acceleration. In this case the problems connected with pulsation of the ventilated cavities can arise [43]

8.2 Balancing the SC-Vehicle

During the SC-vehicle marching motion in the planing regime the vehicle weight must be compensated by the hydrodynamic forces F_{ny} and F_{sy} created respectively by inclining the cavitator on the angle $\delta_z < 0$ and as a result of planning of the vehicle tail along the lower cavity wall, and also, possibly, by deflecting the propulsor thrust vector $\vec{F}_{pr}$ on the angle $\eta_z > 0$ (see Fig. 13). In the steady horizontal motion a sum of the vertical projections of these three forces and the total moment of these forces must be equal to zero.

To determine the balanced values of the angles δ_z^*, η_z^*, and ψ^* one considers projections of the forces on the cavitator F_{ny} and F_{sy} in the body coordinate system (29), and F_{nx0} and F_{ny0} in the flow coordinate system (30). In the steady horizontal motion of the SC-vehicle the following relations must be fulfilled:

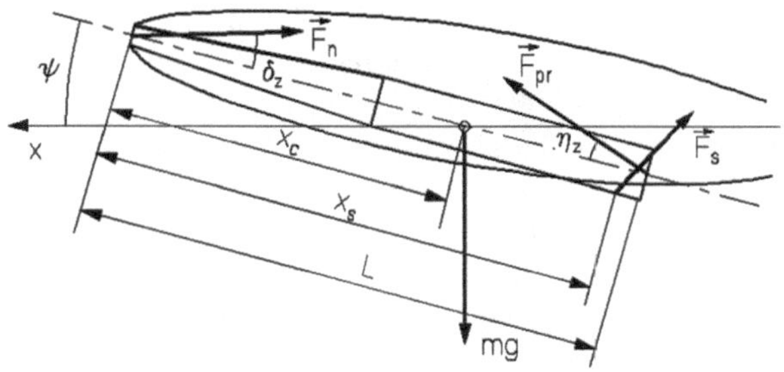

Fig. 13 Scheme of forces acting onto SC-vehicle in steady motion

$$F_{pr}\cos(\eta_z+\psi)-F_{nx0}-F_{sx}\cos\psi=0, \tag{31}$$

$$F_{ny0}+F_{sy}\cos\psi+F_{pr}\sin(\eta_z+\psi)-mg=0, \tag{32}$$

$$F_{ny}x_c-F_{sy}(x_s-x_c)-F_{pr}\sin\eta_z(L-x_c)=0, \tag{33}$$

where x_s is distance from the cavitator to the point of application of the force F_{sy}. The connection between the projections F_{ny} and F_{ny0} follows from (29) and (30):

$$F_{ny}=F_{ny0}\frac{\sin\delta_z}{\sin(\delta_z+\psi)}. \tag{34}$$

From Eqs. 32 and 33 with taking into consideration (34) one obtains the values of the force projections which are required for balancing the SC-vehicle:

$$F^*_{ny0}=\frac{mg-F_{pr}\cos\eta_z\sin\psi}{1+\dfrac{\sin\delta_z\cos\psi}{\sin(\delta_z+\psi)}\dfrac{x_c}{x_s-x_c}},$$

$$F^*_{sy}=F^*_{ny0}\frac{\sin\delta_z}{\sin(\delta_z+\psi)}\frac{x_c}{x_s-x_c}-F_{pr}\sin\eta_z.$$

If the angle η_z is fixed, the balanced values of the angles δ^*_z and ψ^* are determined numerically with the help of an iteration process. For the SC-vehicle model (Fig. 6) when $\bar{x}_c=0.6$ and $\eta_z=0$ the calculation gives $\delta^*_z=-5.781°$ and $\psi^*=0.334°$. If the angle ψ is fixed, the balanced values of the angles δ^*_z and η^*_z are determined analogously.

8.3 Instability of Longitudinal Motion of SC-Vehicle

It is possible to judge about the SC-vehicle motion stability by the computer simulation results for the balanced starting conditions $\delta_z(0)=\delta^*_z$, $\eta_z(0)=\eta^*_z$, $\psi(0)=\psi^*$. The performed calculations have shown that uncontrolled motion of the SC-vehicle is unstable [8]. The instability appears in spontaneous arising and increasing the vehicle oscillations by the pitch angle ψ and in increasing the deviation y_c of the vehicle mass center from the horizontal path. The vehicle behavior after the stability loss strongly depends on its mass center position x_c. The example of development of the SC-vehicle oscillation on the distance 500 m when $\bar{x}_c = 0.6$ is shown in Fig. 14. It is designated $\bar{x} = x/L$ in the figure; $\bar{\omega} = \omega_z L/V_0$, the magnitudes of ψ are in radians.

Amplitude of the angular oscillation of the vehicle grows until the opposite cavity wall begins to confine them. After this the vehicle continues to oscillate within the cavity with the approximately fixed frequency f, which may be named as the natural frequency of the SC-vehicle. In this case the deviation y and the mean value of the angle ψ grow monotonically. A dependence of the reduced natural frequency $k = 2\pi fL/V_0$ on the dimensionless mass center position $\bar{x}_c$ is shown in Fig. 15. The deviations of the vehicle mass center y in meters at the end of distance 300 m are plotted there as well.

As is obvious, the natural frequency essentially depends on the mass center position. In this case the value $\bar{x}_c = 0.5$ approximately divides the function

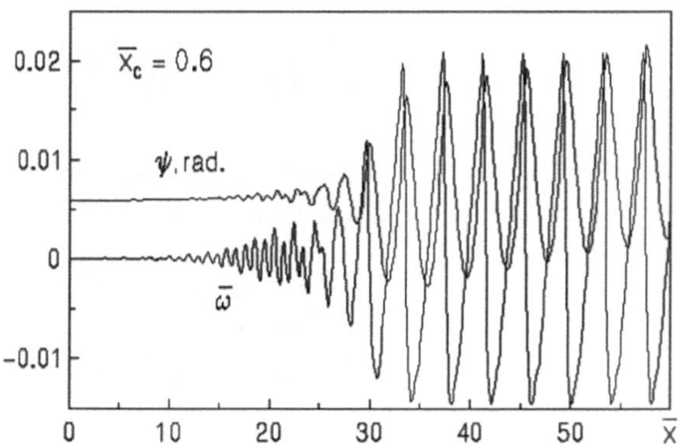

Fig. 14 Instability of the longitudinal motion of the SC-vehicle

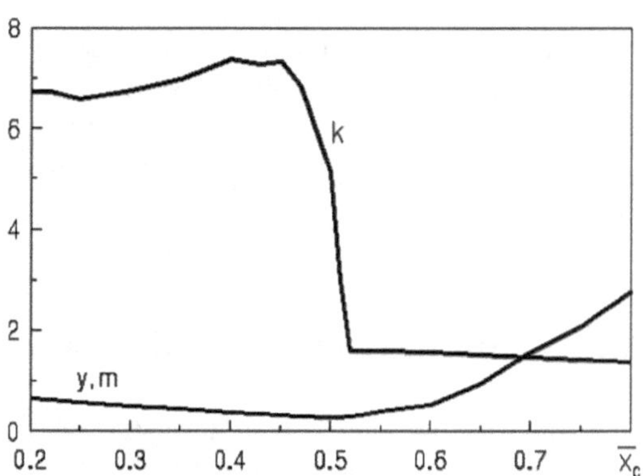

Fig. 15 Influence of x_c on the natural frequencies of the uncontrolled SC-vehicle

variation domain onto two zones. When $\bar{x}_c < 0.5$, the vehicle performs complex oscillation with the high frequency. When $\bar{x}_c > 0.5$, the angular oscillations of the vehicle are close to the simple periodic ones, their frequency decreases, and the mass center deviation increases. The calculations show that it is possible for any values of x_c to determine such distance x that the deviation of the SC-vehicle mass center exceeds any value given beforehand. When $\eta_z = 0$, the uncontrolled SC-vehicle usually exits on the water surface. Thus, introduction of the artificial stabilization is necessary to maintain the horizontal motion of the SC-vehicle.

8.4 Stabilization of SC-Vehicle Motion by δ-Control

One checks an effectiveness of active stabilization of the SC-vehicle motion by automatic regulation of the cavitator inclination angle (δ-control). The vehicle deviation from the balanced position is defined by two time functions $y(t)$ and $\psi(t) - \psi^*$. One accepts a law of the automatic regulating the cavitator inclination angle δ_z in the form:

$$\delta_z(t) = \delta_z^* + k_1\bar{y}(t - t_1) + k_2[\psi(t - t_1) - \psi^*] + k_3\bar{\omega}(t - t_1), \tag{35}$$

where k_1, k_2, k_3 are the non-negative feedback coefficients (the transfer ratios of controller); $t_1 > 0$ is the lag time of the actuator device reaction. It is designated $\bar{y} = y/L$; $\bar{\omega} = \dot{\psi}L/V_m$ in formula (35).

The analogous linear feedback laws were considered in papers [26, 31, 32] and in a number of other papers. The methods of the linear theory of automatic control systems are used there for synthesis of the SC-vehicle automatic control system. However, the set of equations of the SC-vehicle dynamics, formally, does not assume linearization without loss of its key properties. The described mathematical model allows the optimal magnitudes of the transfer-ratios k_1, k_2, and k_3 to be determined by computer simulation of the SC-vehicle motion basing on the complete set of the nonlinear equations [8].

Figure 16 gives an example of calculation of the dependences $\bar{y}(\bar{x})$, $\psi(\bar{x})$, and $\eta_z(\bar{x})$ at the δ-stabilization of the SC-vehicle motion on the distance 1 km ($k_1 = 2$, $k_2 = 5$, $k_3 = 0$, $\bar{x}_c = 0.6$; the angle magnitudes are in degrees). The automatic δ-control (35) transforms unsteady oscillation of the vehicle with the increasing mean magnitude of ψ into the steady oscillation. In this case the operating control (cavitator) performs steady oscillation by the angle δ_z with the same frequency 6.55 Hz.

One notes that measuring the SC-vehicle deviation on the depth with an accuracy required for the automatic control system is a problematic task, whereas the pitch angle deviation is easily measured. In practice, the automatic control system purpose is prohibiting the depth deviation of the SC-vehicle beyond the bounds of the established interval on the established distance. The computer simulation shows

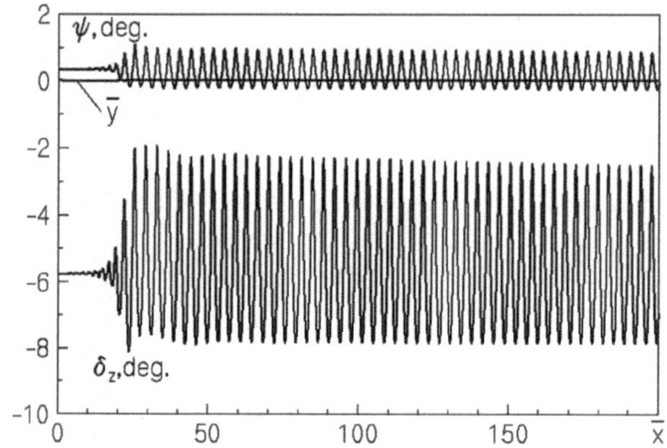

Fig. 16 Changing the motion parameters of the SC-vehicle at the depth δ-control

that using the single-circuit automatic control system by deviation of ψ usually is sufficient for this purpose.

8.5 Stabilization of SC-Vehicle Motion by η-Control

Automatic regulating the deflection of the propulsor thrust vector η_z (η-control) is another principal possibility of active stabilization of the SC-vehicle motion. When inclining the cavitator and deflecting the thrust vector on the angles of equal sign, the forces having opposite direction arise. Therefore, one accepts the law of automatic η-control in the form:

$$\eta_z(t) = \eta_z^* - a_1\bar{y}(t - t_1) - a_2[\psi(t - t_1) - \psi^*] - a_3\bar{\omega}(t - t_1), \tag{36}$$

where a_1, a_2, a_3 are nonnegative feedback coefficients. The calculations showed that at the η-stabilization a character of changing the controlling angle η_z strongly depends on the mass center position x_c.

Figure 17 shows a result of the η-stabilization of the SC-vehicle on distance 1 km ($a_1 = 1, a_2 = 0.5, a_3 = 0, \bar{x}_c = 0.6$). In this case the mean frequency of oscillation of the angles η_z and ψ is equal to 5.93 Hz. Comparing Figs. 16 and 17, it is possible to conclude that the automatic η-control for the SC-vehicle depth stabilization has practically the same effectiveness as the δ-control.

9 Equations of Lateral Motion of SC-Vehicles

The lateral motion problem arises in connection with investigation of the SC-vehicle maneuverability on course. The lateral vehicle motion consists in revolution around the axes O_1x_1 and O_1y_1 and in transition in direction to the axis O_1z_1. If the undisturbed motion is longitudinal, and the deflections are small, then the lateral motion of the SC-body may be approximately considered independently on the

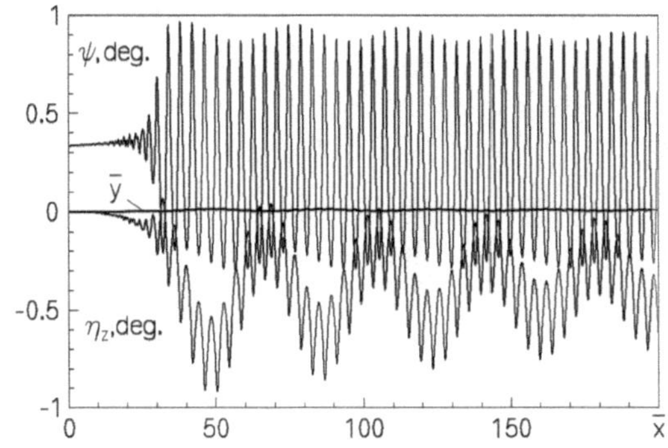

Fig. 17 Changing the motion parameters of the SC-vehicle at the depth η-control

longitudinal motion [36]. When formulating the problem on the SC-vehicle lateral motion one accepts that the longitudinal motion is steady, i.e. $y = 0$, $\alpha = \psi = \psi^*$, $V = V_0$.

Then six Eqs. 3–5, 7, 8, and 12 remain from the general set of dynamic equations (1)–(12). One should assign $\omega_z = 0$, $d\psi/dt = 0$ there. Passing to differentiating with respect to the absolute coordinate x and taking into account the relations (13), we obtain the calculation set of six differential equations of the SC-vehicle lateral motion:

$$\cos(\varphi - \beta)\frac{dV_{z1}}{dx} = \omega_x \sin\psi^* \cos\beta + \omega_y \cos\psi^* \cos\beta + \frac{1}{mV_0}F_{z1}, \tag{37}$$

$$V_0 \cos(\varphi - \beta)\frac{d\omega_x}{dx} = \frac{1}{I_x}M_{x1}, \tag{38}$$

$$V_0 \cos(\varphi - \beta)\frac{d\omega_y}{dx} = \frac{1}{I_y}M_{y1}, \tag{39}$$

$$V_0 \cos(\varphi - \beta)\frac{d\theta}{dx} = \omega_x - \frac{\sin\psi^*}{\cos\psi^* \cos\theta}\omega_y, \tag{40}$$

$$V_0 \cos(\varphi - \beta)\frac{d\varphi}{dx} = \frac{1}{\cos\psi^* \cos\theta}\omega_y, \tag{41}$$

$$\frac{dz}{dx} = -\mathrm{tg}(\varphi - \beta), \tag{42}$$

where $\beta = \arcsin\frac{V_{z1}}{V_0}$. For any x the elapsed time can be calculated by the formula:

$$t(x) = \int_0^x \frac{ds}{V_0 \cos(\varphi - \beta)}.$$

In the case of the lateral motion, the right parts of Eqs. 37–39 and relations (18), (19) take the following form [9]:

$$F_{z1} = F_{nz} + F_{sz} + mg\cos\psi^*\sin\theta - F_{pr}\sin\eta_y,$$

$$M_{x1} = -mg\cos\psi^*\sin\theta\, h_M,$$

$$M_{y1} = -F_{nz}x_{mc} - F_{sz}(x_s - x_c) + F_{prz}(L - x_c),$$

$$F_{nz} = -F_{x0}c_{11}^{(n)}\sin\delta_y\cos\delta_z^*,$$

$$F_{nz0} = -F_{x0}c_{11}^{(n)}c_{13},$$

$$c_{11}^{(n)} = \cos\delta_y\cos\delta_z^*\cos\psi^*\cos\beta_n - \\ - \sin\delta_z^*\sin\psi^*\cos\beta_n - \sin\delta_y\cos\delta_z^*\sin\beta_n,$$

$$c_{13} = -\cos\delta_y\cos\delta_z^*\cos\psi^*\sin\beta + \\ + \sin\delta_z^*\sin\psi^*\sin\beta - \sin\delta_y\cos\delta_z^*\cos\beta.$$

At intersection of the body contour and the cavity contour in the body section by the horizontal plane $O_1x_gz_g$, the lateral planing force is calculated by the formula

$$F_{sz} = \rho\pi R_s^2 V_0\left[V_1\frac{\bar{h}(2+\bar{h})}{(1+\bar{h})^2} + V_2\frac{2\bar{h}}{1+\bar{h}}\right]\cos\theta.$$

The set of differential equations (37)–(42) is integrated numerically for the initial conditions:

$$V_{z1}(0) = V_0\sin\varphi_0,\quad \omega_x(0) = \omega_{x0},\quad \omega_y(0) = \omega_{y0},$$
$$\theta(0) = \theta_0,\quad \varphi(0) = \varphi_0,\quad z(0) = 0.$$

Performed calculations of the lateral motion of various SC-vehicles showed that as distinct from the longitudinal motion in this case spontaneous violation of the motion stability does not occur. It means that for any $x > 0$ the SC-vehicle motion remains purely longitudinal if $\delta_y = 0$, $\eta_y = 0$, $\theta_0 = 0$, $\varphi_0 = 0$, $\omega_{x0} = 0$, and $\omega_{y0} = 0$. The course deviation can arise at violation of any of these conditions and also under influence of outer perturbations.

10 Maneuverability of SC-Vehicles

The maneuverability of the underwater vehicle is its ability to perform a circulation with established radius R_t in the horizontal plane or in the vertical plane. A simple analysis shows that owing to the small surface of contact with water the high-speed SC-vehicles are able to perform circulations with the minimal radius on 2–3 orders bigger than in the continuous flow [44]. Therefore, the SC-vehicle maneuverability on course and on depth may be investigated separately in frames of models of the lateral and the longitudinal motion.

10.1 Course-Maneuverability of SC-Vehicle

The SC-vehicle ability to perform the course maneuver may be characterized by a dependence of the vehicle trajectory angle in the horizontal plane $\chi = \varphi - \beta$ on the cavitator inclination angle δ_y or the angle of the thrust vector deflection η_y. Knowing the function $\chi(x)$, one can calculate easily the circulation radius R_{ty} which is equal to the local curvature radius of trajectory $z = z(x)$:

$$R_{ty}(x) = \frac{(1 + \mathrm{tg}^2\chi(x))^{\frac{3}{2}}}{\mathrm{tg}'\chi(x)}.$$

The calculations showed that the SC-vehicle behavior is different at the δ-control and the η-control (see more detailed in [9]). At the δ-control the SC-vehicle for any δ_y rapidly reaches the regime of steady oscillations by the yaw angle φ periodically interacting with the opposite lateral cavity walls. In this case the frequency and the amplitude of oscillations of φ practically do not depend on δ_y. At the η-control for non-small η_y the vehicle oscillations by φ touching only one of the lateral cavity walls. In this case the oscillation frequency increases, and amplitude decreases when η_y increases.

Figure 18a,b shows dependencies of deviation of the SC-vehicle mass center $\bar{z} = z/L$ and the mean trajectory angle χ_m at the end of distance 500 m on δ_y (*a*) and on η_y (*b*). As is obvious, at the δ-control these dependencies are monotonic and practically linear. On the contrary, at the η-control they are non-monotonic, and the maximal magnitudes of $\bar{z}$ and χ_m are lower on an order than at the δ-control. All this testifies about ineffectiveness of the η-control of the SC-vehicle motion in comparison with the δ-control. This is explained by that at the δ-control the averaged tail force of planning F_{sz} and the lateral force on the cavitator F_{nz} are directed in one side and give in a sum the resultant centripetal force applied in the vehicle mass center. On the contrary, at the η-control the averaged force F_{sz} is directed against the lateral component of the thrust force and weakens its action.

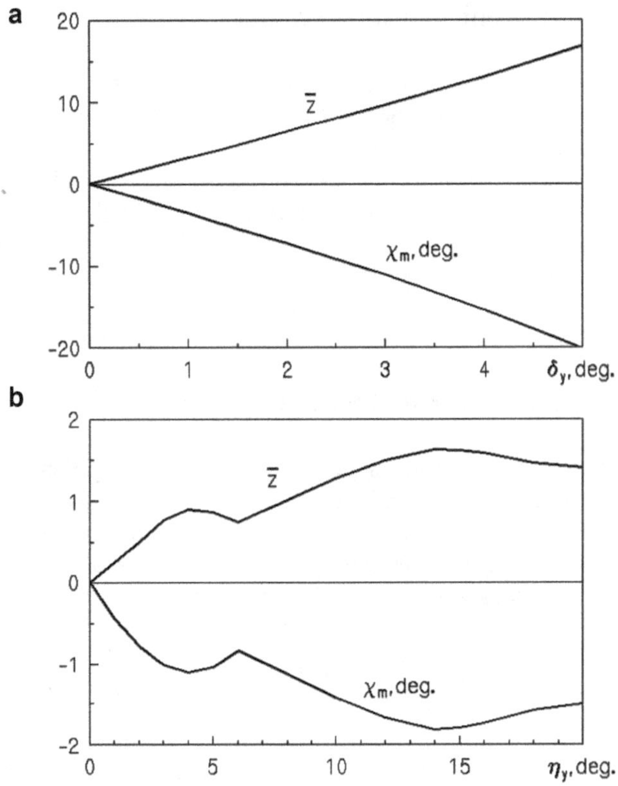

Fig. 18 Deviation $\bar{z}$ and mean trajectory angle χ_m at the SC-vehicle course maneuvering *a* – δ-control; *b* – η-control

Also, we investigated the influence of the starting roll angle $\theta_0 \neq 0$ onto the SC-vehicle trajectory [9]. In this case, if the vehicle metacentric height h_M>0, then the vehicle performs undamped oscillation on the roll angle θ with the amplitude θ_0. The h_M increasing leads to some increasing the frequency of this oscillation.

The vehicle oscillation on the roll causes the cavity oscillation in the horizontal plane, which amplitude increases with decreasing h_M. If in this case the vehicle does not contact with the lateral cavity walls, then variations of $\bar{z}$ and χ_m are insignificant. However, for sufficiently low values of h_M the amplitude of the cavity oscillation increases so that the lateral cavity walls begin to interact with the vehicle. In this case the SC-vehicle oscillation on the yaw angle φ arises, and the mean angle of the trajectory χ_m abruptly increases. In practice with the purpose of the course stabilization of the SC-vehicle motion one should increase the value of h_M, if it is possible, and/or damp the vehicle roll oscillation with the specially designed fins.

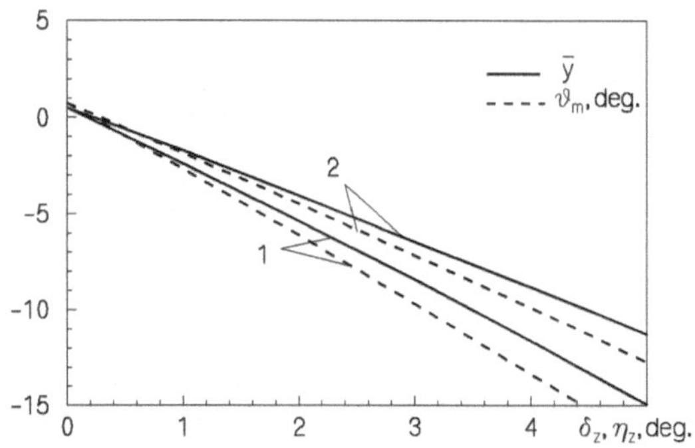

Fig. 19 Deviation $\bar{y}$ and mean trajectory angle ϑ_m at the SC-vehicle depth maneuvering 1 – δ-control; 2 – η-control

10.2 Depth Maneuverability of SC-Vehicles

An ability of the SC-vehicle to perform a maneuver on depth may be characterized by the dependence of the vehicle trajectory angle in the vertical plane $\vartheta = \psi - \alpha$ on the cavitator inclination angle δ_z or on the angle of the thrust vector deflection η_z. Knowing the function $\vartheta(x)$, one can calculate easily the circulation radius R_{tz} which is equal to the local curvature radius of trajectory $y = y(x)$:

$$R_{tz} = \frac{(1 + tg^2\vartheta(x))^{\frac{3}{2}}}{tg'\vartheta(x)}.$$

The specific problem at the SC-vehicle depth maneuvering is caused by varying the hydrostatic water pressure p_∞ and, as a result, varying the cavitation numbers σ. In this case to maintain invariable cavity dimensions it is necessary to regulate the cavity pressure p_c by means of gas-supply and/or to apply the transformer cavitators with variable drag [3].

Figure 19 shows an example of calculation of dependencies of deviation of the SC-vehicle mass center and the mean trajectory angle ϑ_m after flying the distance 500 m on the angle $\delta_z^* - \delta_z$ at the δ-control (curves 1) and on the angle $-\eta_z$ at the η-control (curves 2) ($\bar{x}_c = 0.6$; magnitudes of ϑ_m are in degrees). In the calculations, $\sigma = const$ was accepted at varying the motion depth. One can see, as distinct from the case of the course maneuvering, for the SC-vehicle depth maneuvering the η-control practically has the same effectiveness as the δ-control. This is explained by that at the depth η-control like at the δ-control the variation of the transversal force on the cavitator F_{ny} has the main significance owing to violation of the SC-vehicle balancing.

Conclusions

The particular cases of the longitudinal motion and the lateral motion have been considered basing on the general mathematical model of 3D motion of the SC-body. The problems of the motion stability and optimization for the SC-bodies moving on inertia on an arbitrary angle to the horizon were investigated. The comparative analysis of application of two methods of control of the SC-vehicle motion – by inclining and/or turning the cavitator having two degrees of freedom (δ-control), and by deflection of the propulsor thrust vector (η-control) – has been performed to stabilize the SC-vehicle motion and to maneuver on course and depth. The performed investigations allow the following main conclusions to be made.

1. The motion of the "small" SC-bodies on inertia can be unstable or stable "in the whole" due to periodic interaction of the body with the cavity walls. We have developed the numerical and analytical method of optimization of the SC-body parameters that have the purpose of increasing the motion stability degree and attaining the maximal motion range.
2. The longitudinal motion of the uncontrolled SC-vehicle in the planing within the cavity regime is unstable "in the small". In this case, the automatic regulation of the cavitator inclination angle and/or deflection of the propulsor thrust vector with the linear feedback law permits the problem of practical depth stabilization of the SC-vehicle motion at the specified distance to be solved.
3. The control of the SC-vehicle motion by inclining and/or turning cavitator (δ-control) and the control by deflecting the propulsor thrust vector (η-control) for the motion stabilization and maneuvering are equally effective.
4. Applying the η-control for the course maneuvering the SC-vehicle is ineffective in comparison with the δ-control. Obviously, the same may be said about the control by the tail rudders piercing the cavity walls. One notes that G.V. Logvinovich has pointed onto inefficiency of the "tail control" for the SC-vehicles in the beginning of the 1970s.

References

1. Logvinovich GV. Hydrodynamics of flows with free boundaries. Kiev: Naukova dumka; 1969. 208p. (In Russian). English translation: Halsted Press, 1973.
2. Logvinovich GV. Problems of the theory of slender axisymmetric cavities. Trudy TsAGI. 1976;1797:3–17 (In Russian).
3. Savchenko YuN, Vlasenko YuD, Semenenko VN. Experimental investigations of high-speed cavitation flows. Hidromehanika. 1998;72:103–11. (In Russian). English translation: Int J Fluid Mech Res. 1999;26(3):365–74.
4. Savchenko YuN, Semenenko VN, Putilin SI. Unsteady processes in supercavitation motion of bodies. Prykladna hidromehanika. 1999;1(1):62–80. (In Russian). English translation: Int J Fluid Mech Res. 2000;27(1):109–37.
5. Semenenko VN. Computer simulation of dynamics of supercavitatating bodies. Prykladna Hidromehanika. 2000;2(1):64–9 (In Russian).
6. Nesteruk IG, Semenenko VM. Optimization problems for supercavitation inertial motion of axisymmetric bodies. Prykladna Hidromehanika. 2006;8(1):51–9 (In Ukrainian).

7. Nesteruk IG, Savchenko YuM, Sememenko VM. Optimization of the range for the supercavitation motion on inertia. Dopovidi NAN Ukrainy. 2006;8:57–66 (In Ukrainian).
8. Semenenko VN. Modelling the longitudinal motion of the underwater supercavitating vehicles. Prykladna Hidromehanika. 2010;12(4):81–8 (In Russian).
9. Savchenko YuN, Semenenko VN. On the course maneuvering the underwater supercavitating vehicle. Prykladna Hidromehanika. 2011;13(1):43–50 (In Russian).
10. Savchenko YuN, Semenenko VN, Putilin SI, et al. Designing the high-speed supercavitating vehicles. Proceedings of the 8th International Conference on Fast Sea Transportation (FAST'2005). St. Petersburg; 2005.
11. Savchenko YuN, Semenenko VN, Putilin SI, et al. Some problems of the supercavitating motion management. Sixth International Symposium on Cavitation CAV2006. Wageningen; 2006.
12. Nesteruk IG, Savchenko YuN, Semenenko VN. Achievement of maximal range of supercavitating body inertial motion. Proceedings of the International conference on subsea technologies (SubSeaTECH2007); June 25–28, 2007. St. Petersburg; 2007.
13. Semenenko VN. Some problems of supercavitating vehicle designing. Proceedings of the International Conference on superfast marine vehicles moving above, under and in water surface (SuperFAST'2008); 2–4 July 2008. St. Petersburg; 2008.
14. Savchenko YuN. Experimental investigation of supercavitating motion of bodies. VKI/RTO Special Course on Supercavitation. Von Karman Institute for Fluid Dynamics. Brussels; 2001. (Belgium).
15. Savchenko YuN. Control of supercavitation flow and stability of supercavitating motion of bodies. VKI/RTO Special Course on Supercavitation. Von Karman Institute for Fluid Dynamics. Brussels; 2001. (Belgium).
16. Savchenko YuN. Supercavitating object propulsion. VKI/RTO Special Course on Supercavitation. Von Karman Institute for Fluid Dynamics. Brussels; 2001. (Belgium).
17. Semenenko VN. Artificial supercavitation. Physics and calculation. VKI/RTO Special Course on Supercavitation. Von Karman Institute for Fluid Dynamics, Brussels; 2001. (Belgium).
18. Semenenko VN. Dynamic processes of supercavitation and computer simulation. VKI/RTO Special Course on Supercavitation. Von Karman Institute for Fluid Dynamics. Brussels; 2001. (Belgium).
19. Logvinovich GV. Some problems of planing. Trudy TsAGI. 1980;2052:3–12 (In Russian).
20. Savchenko YuN. Investigations of supercavitation flows. Prykladna Hidromehanika. 2007;9(2–3):150–8 (In Russian).
21. Kulkarni SS, Pratar R. Studies on the dynamics of a supercavitating projectiles. Appl Math Model. 2000;24(2):113–29.
22. Abe A, Katayama M, Saito T, Takayama K. Numerical simulation on supercavitation and jawing of a supersonic projectile traveling in water. Symposium on Interdisciplinary Shock Wave Research. Sendai; March 22–24, 2004.
23. Lindau JW, Kunz RF, Mulherin JM, Dreyer JJ, Stinebring DR. Fully coupled, 6-DOF to URANS, modelling of cavitating flows around a supercavitating vehicle. Fifth International Symposium on Cavitation CAV2006. Osaka; 2003
24. Kirschner I, Rosenthal BJ, Uhlman JS. Simplified dynamical systems analysis of supercavitating high-speed bodies. Fifth International Symposium on Cavitation CAV2006. Osaka; 2003.
25. Kirschner IN, Kring DC, Stokes AW, Fine NE, Uhlman JS. Control strategies for supercavitating vehicles. J Vib Control. 2002;8:219–42.
26. Dzielski J, Kurdila A. A benchmark control problem for supercavitating vehicles and an initial investigation of solution. J Vib Control. 2003;19(7):791–804.
27. Ruzzene M, Kamada R, Botasso CL, Scorceletti F. Trajectory optimization strategies for supercavitating undervater vehicles. J Vib Control. 2008;14(5):611–44.
28. Botasso CL, Scorceletti F. Trajectory optimization for DDE models of supercavitating undervater vehicles. Online preprint; 2008. http://www.aero.polimi.it/~bottasso/.

29. Balas GJ, Bokor J, Vanek B, Arndt REA. Control of high-speed undervater vehicles. In: B.A. Francis et al. (eds.) Control of uncertain systems. Berlin/Heidelberg: Springer-Verlag; 2006. p. 25–44.
30. Vanek B, Bokor J, Balas GJ, Arndt REA. Longitudinal motion control of a high-speed supercavitation vehicle. J Vib Control. 2007;13(2):159–84.
31. Lin G, Balachndran B, Abed EH. Nonlinear dynamics and bifurcations of a supercavitating vehicle. IEEE J Ocean Eng. 2007;32(4):753–61.
32. Lin G, Balachandran B, Abed E. Dynamics and control of supercavitating vehicles. J Dynam Syst Meas Control. 2008;130(2):1–11.
33. Li DJ, Zhang YW, Luo K, Dang JJ. Motion control of underwater supercavitating projectiles in vertical plane. Mod Appl Sci. 2009;3(2):60–5.
34. Li DJ, Luo K, Zhang YW, Dang JJ. Studies on fixed-depth control of supercavitating vehicles. Acta Automatica Sinica. 2010;36(3):421–6.
35. Polyahov NN, Zegzhda SA, Yushkov MP. Theoretical mechanics. Moscow: Vysshaya shkola; 2000 (In Russian).
36. Lukomskiy YuA, Chugunov VS. Systems of controls of marine vehicles. Leningrad: Sudostroenie; 1988 (In Russian).
37. Logvinovich GV, Serebryakov VV. On the methods of calculating a shape of the slender axisymmetric cavities. Hidromehanika. 1975;32:47–54 (In Russian).
38. Buyvol VN. Slender cavities in flows with perturbations. Kiev: Naukova Dumka; 1980 (In Russian).
39. Vasin AD, Paryshev EV. Immersion of a cylinder in liquid through a cylindrical free surface. Izvestiya AN SSSR, Mehanika zhidkosti i gasa. 2001;2:3–12 (In Russian).
40. Schlichting H. Boundary layer theory. New York: McGraw-Hill; 1961.
41. Putilin SI. Some features of a supercavitating model dynamics. Prykladna hidromehanika. 2000;2(3):65–74 (In Russian). English translation: Int J Fluid Mech Res. 2001;28(5):631–43.
42. Nesteruk IG, Semenenko VM. Problems of optimization of a range of the supercavitation motion on inertia with a fixed final depth. Prykladna Hidromehanika. 2006;8(4):33–42 (In Ukrainian).
43. Semenenko VN. Computer simulation of pulsations of ventialed supecavities. Hidromehanika. 1997;71:110–8. (In Russian). English translation: Int J Fluid Mech Res. 1996;23(3 & 4): 302–12.
44. Savchenko YuN. Perspectives of the supercavitation flow applications. Proceedings of the International Conference on Superfast Marine Vehicles Moving Above, Under and in Water Surface (SuperFAST'2008); 2–4 July 2008. St. Petersburg; 2008.

Water Entry of Thin Hydrofoils

A. G. Terentiev

Abstract
First order methods conformably to entry of thin foils in incompressible fluids are considered. Two of them have been based on complex variables and the speed function as well as the partial time-derivative are obtained, another two methods are based on the time depended Green function and Fourier transformation, respectively. For examples, an entry of inclined wedges with strait sides into weightless and also into gravity medium are presented.

1 Introduction

Penetration of a body into a fluid is one of complicated problems in the hydrodynamics. Only a few cases of this problem have been studied, limited mostly to symmetric entry [1–4]. A non-symmetric but perpendicular entry of thin foils was studied by Yim [5]. Two approaches to solving inclined penetration of thin foils have been considered by Terentiev [6, 7] – one is based on complex velocity, another on the partial time-derivative of complex potential. These methods are presented below.

2 Complex Velocity Approach

Consider the motion of a partially submerged slender body (Fig. 1a). Let the abscissa axis be parallel to principal velocity $\boldsymbol{U}$ of the body so that the y-component of speed, v, is small. The linear value problem is formulated on the half plane with

A.G. Terentiev (✉)
Professor of Cheboksary Polytechnic Institute of Moscow State Open University, Cheboksary, Russia
e-mail: agterent@rambler.ru

I. Nesteruk (ed.), *Supercavitation*,
DOI 10.1007/978-3-642-23656-3_10,

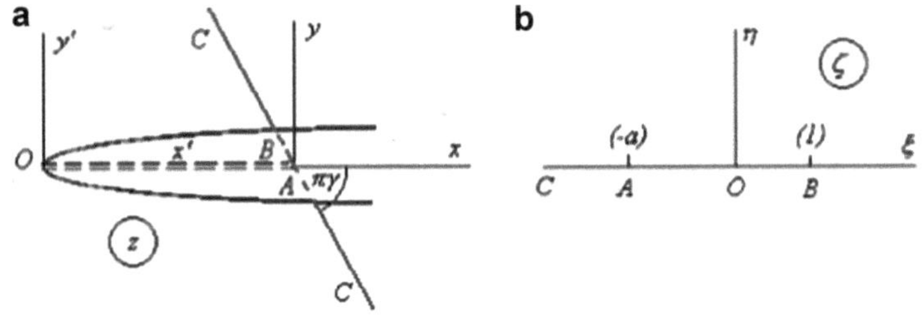

Fig. 1 Inclined entry in fluid: (**a**) – flow in the z-plane; (**b**) – the parametric ζ-plane

an inclined slit (the definitional domain in Fig. 1a is marked by the dashed line). A linear kinematical condition on the wetted surface of the body is imposed as:

$$\mathrm{Im}\left(\frac{\partial w}{\partial z}\right) = g(x,t),\ x \in (l,0),\ y = \pm 0, \tag{1}$$

where $g(z,t)$ is described by the given motion of the body. The dynamic condition on the free boundary (inclined dashed line) is described by

$$\mathrm{Re}\left(\frac{\partial w}{\partial z}e^{-i\pi\gamma}\right) = 0 \text{ on } \mathrm{Im}\left(ze^{\pi\gamma i}\right) = 0. \tag{2}$$

Besides, a solution of the mixed value problem should possess at the leading edge a singularity of the type

$$\partial w/\partial z \approx (z-l)^{-1/2}, \tag{3}$$

and a zero of the second kind at infinity

$$\partial w/\partial z \approx z^{-2}, \quad z \to \infty. \tag{4}$$

The mixed value problem solution can be determined in parametric form by mapping conformably the definitional domain in the z-plane onto the upper half-plane of the auxiliary ζ -plane, Fig. 1b. The transformation is:

$$z == l(1-\zeta)^{\gamma}\left(1+\frac{\zeta}{a}\right)^{1-\gamma}, \quad a = \frac{1-\gamma}{\gamma}. \tag{5}$$

Instead of function $w_z(z,t)$, it is more convenient to solve another value problem of the partial derivative $\partial w/\partial \zeta = w_\zeta(\zeta,t)$, which satisfies the following conditions:

$$\mathrm{Re}\ w_\zeta = 0 \text{ on } \xi \in (-\infty,-a) \cup (1,\infty),\ \eta = 0, \tag{6}$$

$$\mathrm{Im}\ w_\zeta = g(x(\xi,t),t)x_\xi(\xi,t) \text{ on } \xi \in [-a,\ 1]. \tag{7}$$

Multiplying by function, $g(\zeta) = \sqrt{(1-\zeta)(a+\zeta)}$, the mixed value problem is reduced to Schwarz problem with boundary condition,

$$\mathrm{Im}[w_\zeta g(\zeta)] = \begin{cases} f(x(\xi,t)x_\xi(\xi)g(\xi), & \xi \in [-a,\ 1],\ \eta = 0, \\ 0 & \xi \in (-\infty,-a) \cup (1,\ \infty),\ \eta = 0\,. \end{cases} \tag{8}$$

Thus a solution of the value problem is of the form

$$w_\zeta = \frac{la^{\gamma-1}}{\pi\sqrt{(1-\zeta)(a+\zeta)}} \int\limits_{-a}^{1} \left(\frac{a+\tau}{1-\tau}\right)^{\frac{1-2\gamma}{2}} \frac{g\tau d\tau}{\tau-\zeta}. \tag{9}$$

Complex velocity, $w_z = w_\zeta(\zeta,t)/z_\zeta(\zeta,t)$, is a function of variable ζ and time t together with Eq. 4, they represent the value problem solution presented in Eqs. 1–4. Since variable ζ is a function of z and t, it is quite complicated to obtain complex potential $w(z,t)$ and its partial time derivative $w_t(z,t) = \partial w(z,t)/\partial t$. But on the surface of the body, potential $\varphi(\xi,t)$ can be determined integrating equation (9) over interval $(-a,\ \xi)$, and then pressure can be expressed in the form:

$$p = -\rho\left(\frac{\partial\varphi(\xi,t)}{\partial t} + \frac{\partial\xi(x,t)}{\partial t}\frac{\partial\varphi(\xi,t)}{\partial\xi}\right), \tag{10}$$

where

$$\frac{\partial\xi(x,t)}{\partial t} = \frac{\dot{l}(1-\xi)(a+\xi)}{l\xi}. \tag{11}$$

The potential on $[-a,\ 1]$ is obtained from (9) by integral

$$\varphi(\xi,t) = \int\limits_{-a}^{\xi} \mathrm{Re}\left\{\frac{\partial w}{\partial\zeta}\right\}_{\zeta=\xi} d\xi = \frac{l}{\pi a^{1-\gamma}} \int f(\tau)I(\tau,\xi)\left(\frac{a+\tau}{1-\tau}\right)^{(1-2\gamma)/\gamma} \tau\, d\tau, \tag{12}$$

where $I(\tau,\xi) = \int\limits_{-a}^{\xi} \frac{d\xi}{\sqrt{(a+\xi)(1-\xi)}(\tau-\xi)} = -\frac{\ln|G(\tau,\xi)|}{\sqrt{(a+\tau)(1-\tau)}}$,

$$G(\tau,\xi) = \ln\left|\frac{\sqrt{(a+\xi)(1-\tau)} - \sqrt{(a+\tau)(1-\xi)}}{\sqrt{(a+\xi)(1-\tau)} + \sqrt{(a+\tau)(1-\xi)}}\right|.$$

Integrating the pressure along the wetted part of the foil, one can calculated force and torque. The transverse force is equal to

$$Y = \int_{-1}^{0} (p_1 - p_2)dx = -\int_{-a}^{1} (p - p_0)\frac{\partial x(\xi, t)}{\partial \xi} d\xi. \tag{13}$$

Since $\frac{\partial \xi(x,t)}{\partial t}\frac{\partial x(\xi,t)}{\partial \xi} = -\frac{\dot{l}}{l}x(\xi,t)$, the latter integral may be rewritten in the form

$$\begin{aligned} Y = &-\frac{\rho}{\pi a^{1-\gamma}} \int_{-a}^{1}\int_{-a}^{1} [2\dot{l}g(\tau) + l\dot{g}(\tau)]\frac{\partial x(\xi,t)}{\partial \xi} \\ &\times \frac{\ln|G(\tau,\xi)|}{(a+\tau)^{\gamma}(1-\tau)^{1-\gamma}}\tau\, d\xi d\tau, \end{aligned} \tag{14}$$

where $\dot{l} = dl/dt = -U$ is the speed of the moving body, $\dot{g}(\tau) = \partial g/\partial t$. The integral with respect to variable ξ can be found analytically [6] so the transverse force is expressed as

$$Y = \frac{\rho l}{2}\int_{-a}^{1}\left(\dot{l}g(\tau) + l\frac{g(\tau)}{2}\right)L(\tau)d\tau, \tag{15}$$

where $L(\tau) = \frac{4\tau}{a^{2(1-\gamma)}\cos\pi\gamma}\begin{bmatrix} \left(\frac{a+\tau}{1-\tau}\right)^{1-2\gamma}\sin\pi\gamma - \\ -\left(\frac{a+\tau}{1-\tau}\right)^{(1-2\gamma)/2} \end{bmatrix}$.

Function $L(\tau)$ has uncertainty as the inclination angle $\gamma \to 1/2$; its limit is as

$$L(\tau) = \frac{4\tau}{\pi}\ln\frac{1+\tau}{1-\tau}, \quad \gamma = \frac{1}{2}. \tag{16}$$

Longitudinal force consists of concentrated force at the leading edge $O(x = -l)$ and of longitudinal projection of pressure on the real axis [6], $X = X_c + X_p$, where

$$\begin{aligned} X_c &= -\overline{\frac{i\rho}{4}\oint_{\zeta=0}\frac{w_\zeta^2}{z_\zeta}d\zeta} \\ &= -\frac{\rho l}{2\pi a^{2(1-\gamma)}}\left[\int_{-a}^{1} g(\tau)\left(\frac{a+\tau}{1-\tau}\right)^{(1-2\gamma)/2}d\tau\right]^2; \end{aligned} \tag{17}$$

$$X_p = \frac{\rho l}{2} \int_{-a}^{1} \int_{-a}^{1} q(\tau, \xi) R(\tau, \xi) d\tau \, d\xi;$$

$$q(\tau, \xi) = [l\dot{g}(\tau) + \dot{l}g(\tau)]\beta(\xi); \tag{18}$$

$$R(\tau, \xi) = \frac{2\tau\xi \ln|G(\tau, \xi)|}{\pi a^{2(1-\gamma)}(a+\tau)^{\gamma}(1-\tau)^{1-\gamma}(a+\xi)^{\gamma}(1-\xi)^{1-\gamma}}.$$

The torque relative to the origin of the fixed coordinates is determined by integral

$$M = -\int_{-a}^{1} (p - p_0) x(\xi, t) \frac{\partial x(\xi, t)}{\partial \xi} d\xi,$$

which may be expressed as the transverse force by single integral:

$$M = -\frac{\rho l^2}{2} \int_{-a}^{1} [\dot{l}g(\tau) + \frac{l}{3}\dot{g}(\tau)] K(\tau) d\tau,$$

$$K(\tau) = \frac{3\tau}{a^{3(1-\gamma)} \cos 2\pi\gamma} \left[(a+\tau)^{2-3\gamma}(1-\tau)^{3\gamma-1} \sin 2\pi\gamma - \right.$$

$$\left. - \left(\tau + \frac{3(1-2\gamma)}{2\gamma} \right) \left(\frac{a+\tau}{1-\tau} \right)^{(1-2\gamma)/2} \right]. \tag{19}$$

For $\gamma \to 1/2$ the function approaches the limit

$$K(\tau) = \frac{\tau}{\pi 3^{5/4}} \left[(3+\tau) \ln \frac{3+\tau}{1-\tau} - 4 \right] \left(\frac{3+\tau}{1-\tau} \right)^{1/4}. \tag{20}$$

3 Entry of a Wedge With Straight Sides

Consider forward movement of the wedge with speed $U = -\dot{l}$. Let angles α_1 and α_2 stand for the upper and lower sides of inclination angle to the x-axis, respectively,

$$g(\tau) = U\alpha_1, \tau \in (0,\ 1), \quad g(\tau) = U\alpha_2, \tau \in (-a,\ 0). \tag{21}$$

Then we have following expressions for the transverse force, longitudinal force, leading edge force and the moment respectively:

$$
\begin{aligned}
Y &= \frac{\rho l}{2}\left(\frac{l\dot{U}}{2} - U^2\right)(\alpha_1 I_1 + \alpha_2 I_2),\\
X &= \frac{\rho l}{2}(U^2 - l\dot{U})\left(\frac{\alpha_1^2 I_1 + \alpha_2^2 I_2}{2} + (\alpha_1 - \alpha_2)^2 I_3\right),\\
X_c &= -\frac{\rho U^2 l}{2}(\alpha_1 I_4 + \alpha_2 I_5)^2,\\
M &= \frac{\rho l^2}{2}\left(U^2 - \frac{\dot{U}l}{3}\right)(\alpha_1 I_6 + \alpha_2 I_7),
\end{aligned}
\tag{22}
$$

Here

$$
I_1 = \int_0^1 L(\tau)d\tau, \; I_3 = \int_{-a}^{0}\int_0^1 R(\tau,\xi)d\tau d\xi,
$$

$$
I_4 = \frac{1}{\sqrt{\pi}a^{1-\gamma}}\int_0^1 \left(\frac{a+\tau}{1-\tau}\right)^{(1-2\gamma)/2} d\tau, \; I_6 = \int_0^1 K(\tau)d\tau.
$$

Factors I_2, I_5 and I_7 are calculated by the same integrals as I_1, I_4 and I_6, respectively, but integrating should be on interval $(-a,\ 0)$. It can be shown using beta-functions that the sums are expressed as:

$$
\begin{aligned}
I_1 + I_2 &= 2(I_4 + I_5)^2 = \frac{2\pi}{a^{2(1-\gamma)}\cos^2\pi\gamma}\left(\frac{1-2\gamma}{2\gamma}\right)^2,\\
I_6 + I_7 &= \frac{\pi(1-2\gamma)(16\gamma^2 - 16\gamma + 3)}{8a^{3(1-\gamma)}\gamma^3\cos\pi\gamma\cos 2\pi\gamma}\ .
\end{aligned}
\tag{23}
$$

All integrals for coefficients I_k can be calculated numerically. They can also be expressed by hypergeometric functions $F(\alpha,\ \beta;\ \gamma; x\, z)$ [8]:

$$
I_1 = \frac{4}{(1-\gamma)^{2(1-\gamma)}\cos\pi\gamma}\left\{\begin{array}{l} \dfrac{2\gamma^{(1-2\gamma)/2}}{1+2\gamma}[(1-\gamma)F_1 - F_2] + \\ + \dfrac{\sin\pi\gamma}{2\gamma}[F_4 - (1-\gamma)F_3] \end{array}\right\},
$$

$$
I_4 = \frac{2\sqrt{\gamma}}{\sqrt{\pi}(1+2\gamma)(1-\gamma)^{1-\gamma}}F_4,
$$

$$I_6 = \frac{3}{(1-\gamma)^{3(1-\gamma)}\cos 2\pi\gamma}\left\{\frac{F_6-(1-\gamma)F_5}{3\gamma}\sin 2\pi\gamma\right.$$
$$\left. + \gamma^{(1-4\gamma)/2}\left[\frac{(1-4\gamma)(1-\gamma)}{1+2\gamma}F_1+F_2-\frac{2F_7}{1+2\gamma}\right]\right\},$$

where

$$\begin{aligned}
F_1 &= F(\gamma+0.5,\ \gamma-0.5;\ \gamma+1.5;\ \gamma),\\
F_{2:} &= F(\gamma+0.5,\ \gamma-1.5;\ \gamma+1.5;\ \gamma),\\
F_3 &= F(2\gamma,\ 2\gamma-1;\ 2\gamma+1;\ \gamma),\\
F_4 &= F(2\gamma,\ 2\gamma-2;\ 2\gamma+1;\ \gamma),\\
F_5 &= F(3\gamma,\ 3\gamma-2;\ 3\gamma+1\gamma),\\
F_5 &= F(3\gamma,\ 3\gamma-3;\ 3\gamma+1;\ \gamma),\\
F_7 &= F(\gamma+0.5,\ \gamma-2.5;\ \gamma+1.5;\ \gamma).
\end{aligned}$$

The hyper geometric function may be calculated by the power series expansion

$$F(\alpha,\ \beta;\ \gamma;\ z) = 1+\sum_{k=1}^{\infty} z^k \prod_{m=0}^{k-1}\frac{(\alpha+m)(\beta+m)}{(\gamma+m)(1+m)}. \tag{24}$$

As inclination angle γ approaches zero, coefficients I_1, I_3, I_4, I_6 approach zero but other coefficients approach as follows: $I_2 \to \pi/2, I_5 \to \sqrt{\pi}/2, I_7 \to 3\pi/8$. This case corresponds to glissade of the plate with variable length; the leading edge moves with a speed U but the other edge is fixed.

For perpendicular entry of a wedge as $\gamma = \pi/2$ all integrals in (22) are expressed in simple analytical forms and the hydrodynamic forces and moment are presented as

$$\begin{aligned}
Y &= \rho l(l\dot{U}-2U^2)(\alpha_1+\alpha_2)/\pi,\\
X &= \rho l(U^2-l\dot{U})[(\alpha_1+\alpha_2)^2/2+(\alpha_1-\alpha_2)^2\ln 2]/\pi,\\
X_0 &= -\rho lU^2(\alpha_1+\alpha_2)^2/2\pi,\\
M &= \rho l^2(3U^2-l\dot{U})(\alpha_1+\alpha_2)/6.
\end{aligned} \tag{25}$$

For example, consider an entry of a plate and a wedge with flat sides. Let the plate's slopes are $\alpha_1=\alpha_2=-\alpha$, and wedge's $\alpha_1=-\alpha_2=\alpha$. The following coefficients are introduced:

$$\begin{aligned}
&C_Y = 2Y/\rho U^2 l\alpha,\ \ C_X = 2X/\rho U^2 l\alpha^2,\\
&C_0 = -2X_c/\rho U^2 l\alpha^2,\ C_M = -2M/\rho U^2 l^2\alpha\,.
\end{aligned} \tag{26}$$

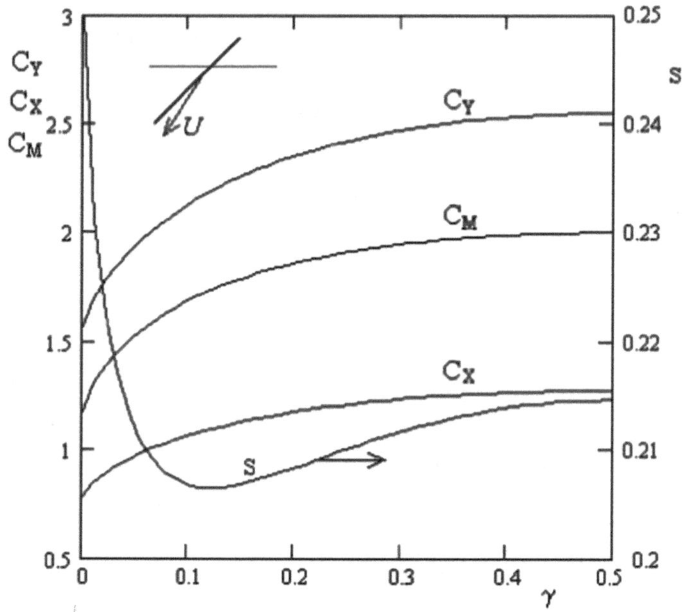

Fig. 2 Entry of a flat plate

The leading edge force and torque are negative in these examples, thus all the coefficients are positive now. Parameter $s = 1 - C_M/C_Y$ is the distance of the pressure center from leading edge related to length l. Dependences of the coefficients are plotted in Figs. 2 and 3.

Note: the scale for dependences of $s(\gamma)$ is shown on the right side on both figures.

One can find that at $\gamma \to 0$ the flat plate and the wedge give the same results: $C_Y \to \pi/2, C_X \to \pi/4, C_M \to 3\pi/8, C_0 \to \pi/4$ and $s \to 1/4$. That is because only one side of the wedge is contiguous to liquid. It should be noted that the lift coefficient is a half of that for a planning plate of a constant length. In another limited case for perpendicular entry, $\gamma \to 1/2$, we have for the flat plate $C_Y \to 8/\pi$, $C_M \to 2$, $C_0 \to 4/\pi$, $s \to 1 - \pi/4$, and for the wedge $C_X \to 8 \ln 2/\pi$; all other coefficients C_Y, C_M and C_0 vanish, so distance s is not meaningful but not zero. It should be noted that the nonlinear problem of perpendicular entry of a wedge with flat sides was considered by using Wagner function in [9].

4 Partial Time-Derivative

The above-mentioned approach is quite difficult for studying many problems associated with unsteady flow with free boundaries due to integration and partial differentiation of very complex functions. It is preferable, as before, to formulate a value problem direct for partial derivative with respect to time, $w_t(z,t) = \partial w(z,t)/\partial t$. This method has been considered by A.G. Terentiev [7] and was

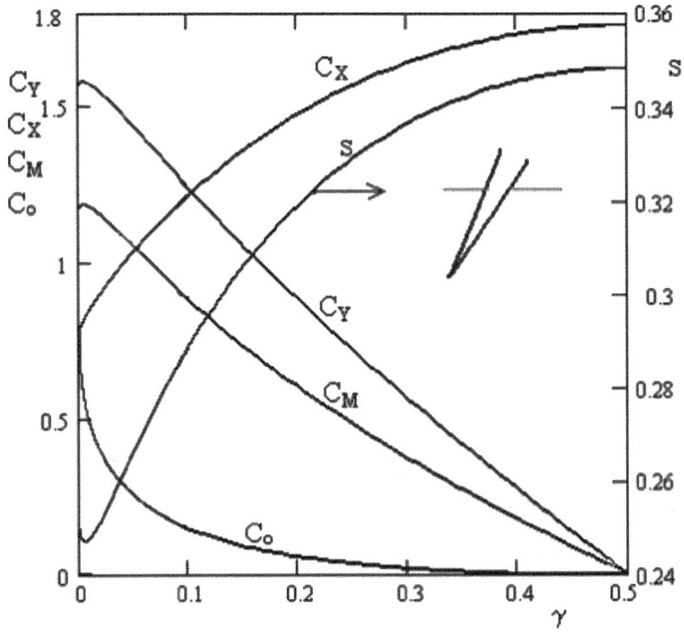

Fig. 3 Entry of a wedge

applied to investigation of many non-steady problems of cavitating as well as non-cavitating flow of foils [4, 10]. Below is an example of using this approach to solve unsteady linear problems of an inclined entry with an attached cavity.

The boundary condition on the moving foil can be rewritten integrating along the x-axis

$$\psi(x,t) = -\int_{s(t)}^{x} g(x,t)dx + T(t). \tag{27}$$

Hence,

$$\psi_t = \dot{s}g(x,t) - \int_{s}^{x} \dot{g}(x,t)dx + \dot{T}(t), \tag{28}$$

where $\dot{T}$ is to be found.

The dynamic condition on the free boundary is

$$\varphi_t = 0. \tag{29}$$

Therefore, function $w_t(z,t)$ should be a solution of a mixed value problem with conditions (28) and (29). A solution of mixed value problem has a certain singularity at the leading edge, but the stream function should be uninterrupted at that point. The condition of continuity of stream function at the leading edge develops an integral equation for unknown function *T(t)*.

5 Entry of a Flat Plate

Consider an inclined entry of a flat plate with a ventilated cavity at a constant velocity (Fig. 4a). On the slit bank *OB*, which has a variable length *l(t)*, the kinematic condition is $\psi_x = -\varphi_y = \dot{l}\alpha$. Due to Eq. 27, the stream function on the plane is of the form:

$$\psi = \dot{l}\alpha(x - l) + T(t). \tag{30}$$

Hence, the boundary conditions for partial time derivative of complex potential on the ξ -axis are as follows:

$$\psi_t = \dot{T} - \dot{l}^2\alpha, \ \xi \in [0,\ 1]; \quad \varphi_t = 0 \text{ outside}(0, 1). \tag{31}$$

Besides, function w_t has a singularity of the form $\zeta^{-1/2}$ at point O, is limited at point B, and approaches zero in infinity. These requirements are met by the function

$$w_t = i(\dot{T} - \dot{l}^2\alpha)\left(1 - \sqrt{\frac{\zeta - 1}{\zeta}}\right). \tag{32}$$

The same function on the ξ-axis determines partial time derivative of the stream function outside (0, 1), so that at any moment the stream function itself can be obtained by integration with respect to time over interval (0, t). Equating with function for *T*(*t*), one obtains an integral equation of the form

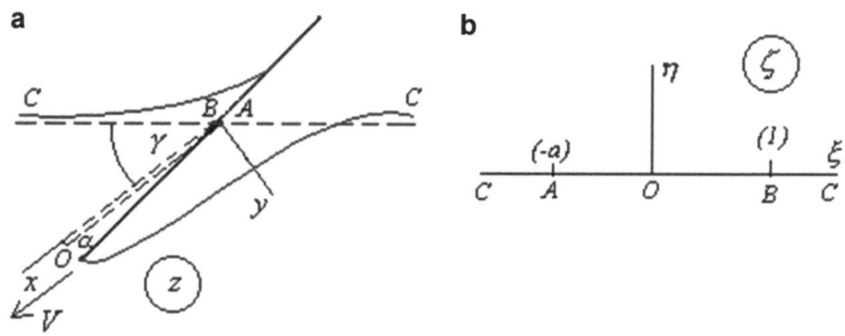

Fig. 4 Entry of plate with ventilated cavity: (**a**) – the z-plane; (**b**) – the parametric ζ-plane

$$\mathrm{Re}\int_{0}^{t}(\dot{T}-\dot{l}^{2}\alpha)\sqrt{\frac{\zeta-1}{\zeta}}\,d\tau=-\dot{l}^{2}t\alpha. \tag{33}$$

Variable ζ is a function of times t and τ as a result of the equality

$$s(t)=s(\tau)(1-\zeta)^{\gamma}\left(1+\frac{\zeta}{a}\right)^{1-\gamma}. \tag{34}$$

For constant speed $V=\dot{l}=const$, the solution of Eq. 33 and all the hydrodynamic characteristics can be expressed in analytic form. Passing to new variable $\sigma=\tau/t$, one can write both last equations as

$$\mathrm{Re}\int_{0}^{1}(\dot{T}-U^{2}\alpha)\sqrt{\frac{\xi-1}{\xi}}d\sigma=-U^{2}\alpha, \tag{35}$$

$$\sigma=(1-\zeta)^{-\chi}\left(1-\frac{\zeta}{a}\right)^{\chi-1}. \tag{36}$$

Since the function $\zeta(t,\ \tau)$ depends only on variable σ, and equality (35) should be satisfied for any value of time, unknown function $\dot{T}-U^{2}\alpha$ should be a constant. Hence,

$$\dot{T}-U^{2}\alpha=-U^{2}\alpha\, I^{-1}, \tag{37}$$

where

$$I=\mathrm{Re}\int_{0}^{1}\sqrt{\frac{\xi-1}{\xi}}\,d\sigma$$

In view of equality (36) the integral in (37) can be submitted through hypergeometrical and beta-functions [8] as

$$\begin{aligned}I&=a^{1-\chi}\times\mathrm{Re}\int_{0}^{\infty}\delta^{1/2}\left(e^{\pi(1-\chi)i}+\delta\right)^{-(\chi+1/2)}\left(\delta+ae^{-\pi\chi i}\right)^{-(2-\chi)}d\delta\\&=a^{1/2}\mathrm{Re}\left\{e^{-\pi i/2}B\left(\frac{3}{2};\ 1\right)F\left(\frac{1}{2}+\gamma,\frac{3}{2};\frac{5}{2};\frac{1}{\gamma}\right)\right\}\\&=\frac{\gamma^{\gamma}}{(1-\gamma)^{1/2}}\left[\begin{array}{l}F\left(\frac{1}{2}+\gamma,\ \gamma-1;\ \gamma;\ \gamma\right)\cos(\pi\gamma)+\\+\frac{1}{2}B\left(\gamma;\ \frac{1}{2}\right)\gamma^{1-\gamma}\end{array}\right].\end{aligned} \tag{38}$$

Resulting force can be found by integrating pressure along the body's surface:

$$Y = \int_0^s p dx = -\frac{\rho U^2 \alpha}{I} \int_0^1 \sqrt{\frac{1-\xi}{\xi}} \frac{\partial x(\xi,t)}{\partial \xi} d\xi =$$
$$= \frac{\rho U^2 s \alpha \gamma^\gamma}{I a^{1-\gamma}} B\left(\gamma + \frac{1}{2}; \frac{3}{2}\right) F\left(\gamma, \gamma + \frac{1}{2}; \gamma + 2; \gamma\right). \tag{39}$$

Whence its coefficient is

$$C_y = \frac{2Y}{\rho U^2 s \alpha} = \frac{\pi F\left(\gamma, \gamma + \frac{1}{2}; \gamma + 2; \gamma\right)}{I\,(1-\gamma)^{1-\gamma}(1+\gamma) B\left(\gamma; \frac{1}{2}\right)}. \tag{40}$$

The torque relative to the origin of the fixed coordinate is calculated similarly. Its coefficient is

$$C_M = \frac{2M}{\rho U^2 s^2 \alpha}$$
$$= \frac{\pi F\left(2\gamma - 1,\ 2\gamma + \frac{1}{2};\ 2\gamma + 2;\ \gamma\right)}{2(1-\gamma)^{2(1-\gamma)}(1+2\gamma) B\left(2\gamma; \frac{1}{2}\right) I}. \tag{41}$$

Depending on values of arguments it is possible to use various representations of special functions [8]. In particular, at small values it is necessary to use other formulas. Namely, Eqs. 40 and 41 for $\gamma \to 0$ yield the above obtained results for coefficients C_X and C_M ($\pi/2$ and $3\pi/8$).

Figure 5 shows the dependencies of the lift coefficient, C_L, and the center of the pressure, x_0, of a plate on the angle incline above. For $\gamma = 1$, the lift coefficient $C_L = \pi/2$ and the center of the pressure $x_0 = 3/4$.

6 Entry of a Wedge of Finite Length

Consider the vertical entry of a wedge of finite length l into a fluid as shown in Fig. 6.

The wedge sides are equal and form angles α_1 and α_2 with the x-axis. The wedge moves at constant velocity U. The boundary conditions are:

$$\psi_t = \begin{cases} -V^2\alpha_1 + \dot{T}, \xi \in (-b, 0), \\ -V^2\alpha_2 + \dot{T}, \xi \in (0, b), \end{cases}$$
$$\varphi_t = 0 \text{ outer interval } (-b, b). \tag{42}$$

The analytical solution of this value problem is

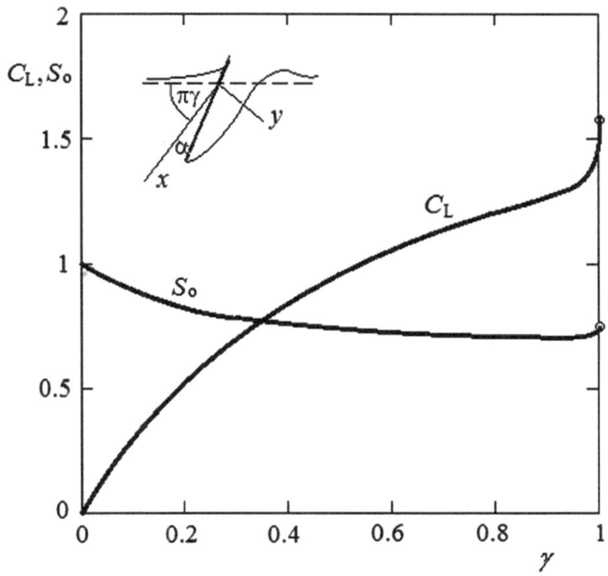

Fig. 5 Dependence of lift coefficient C_L and pressure center s_0 on entry angle γ

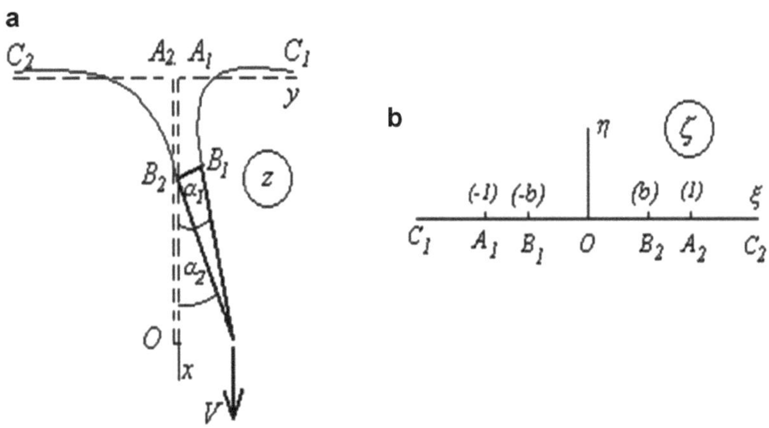

Fig. 6 Vertical entry of a finite-length wedge: (**a**) the z-plane; (**b**) the ζ-plane

$$w_t = -\frac{V^2\beta}{\pi}\ln\frac{\sqrt{\zeta^2 - b^2} + ib}{\zeta} + iV^2\alpha\left(1 - \frac{\sqrt{\zeta^2 - b^2}}{\zeta}\right), \tag{43}$$

where $\alpha = (\alpha_1 + \alpha_2)/2, \ \ \beta = \alpha_2 - \alpha_1$.

Two stages of the wedge entry should be considered: without and with cavity ventilated. The former case was considered above with results in Eq. 25. Nevertheless we shall consider this problem by the method of Volterra's type integral. In this case function (43) could be expressed directly in the z variable as

$$\frac{\partial w(z,t)}{\partial t} = -\frac{U^2\beta}{\pi}\ln\frac{\sqrt{z+s}}{\sqrt{z-s}} + iU^2\alpha\mu(t)\left(1 - \frac{z}{\sqrt{z^2 - s^2}}\right). \tag{44}$$

Whence the stream function on interval $(s, \ \infty)$ is found as

$$\psi(x, \ 0, \ t) = U^2\alpha\int_0^t \mu(\tau)\left(1 - \frac{x}{\sqrt{x^2 - s^2(\tau)}}\right)d\tau, \tag{45}$$

where $\mu = \dot{T}/V^2\alpha - 1$. Substituting $x = s(t)$ and equating unknown function T, we receive the required integral equation:

$$\int_0^t \mu(t)\frac{s(t)}{\sqrt{s^2(t) - s^2(\tau)}}\,d\tau = -t. \tag{46}$$

This equation, as Abel's integral equation, may be solved analytically. But if the wedge moves with a constant velocity, then $s(t) = Ut, \quad s(\tau) = U\tau$, and after substitution $\tau = t\sin\theta$, integral equation (46) takes the form $\int\limits_0^{\pi/2} \mu(t\sin\theta)d\theta$ $= -1$, which has the only solution as $\mu(t) = -2/\pi$.

Consider now a motion of a fully immersed wedge ($s>l$). The stream function on the x-axis outer wedge is determined from (43) by substituting $\zeta = \sqrt{s^2 - z^2}/s$ directly in the x variable as

$$\psi(x, \ 0, \ t) = \\ = U^2\alpha\int_{t_0}^t \mu(\tau)\left(1 - \sqrt{\frac{x^2 - (s(\tau) - l)^2}{x^2 - s^2(\tau)}}\right)d\tau + \psi_0(x). \tag{47}$$

Initial stream function at t_0 can be found from (45)

$$\psi_0(x) = -\frac{2U^2\alpha}{\pi}\left(t_0 - \frac{x}{U}\arcsin\frac{Ut_0}{x}\right).$$

Equating function (47) for $x = Ut, \ s(\tau) = U\tau$, one obtains the integral equation in the form:

$$\int_{t_0}^{t} \mu(\tau)\sqrt{\frac{t^2-(\tau-t_0)^2}{t^2-\tau^2}}d\tau = -t\left(1-\frac{2}{\pi}\arcsin\frac{t_0}{t}\right). \tag{48}$$

This integral equation has an analytical solution only if time approaches to infinity, $\mu \to -1$. This case corresponds to a wedge flow with attached Kirchhoff's cavity in unbounded domain. For any time Eq. 48 can be solved numerically by the iterative method offered above.

Hydrodynamic forces and torque are calculated as usually by integration of pressure along the wedge. Skipping manipulations, the resulting formulas are:
Lift (transverse force)

$$\begin{aligned} Y &= -2\rho U^2\alpha\mu(t)\int_{s-l}^{s}\sqrt{\frac{x^2-(s-l)^2}{x^2-s^2}}dx \\ &= -2\rho U^2\alpha\mu(t)\left[E(b)-\left(1-b^2\right)K(b)\right]; \end{aligned} \tag{49}$$

Concentrated force at the leading edge

$$X_0 = \frac{i\rho}{2U^2}\oint_{z=s}\overline{\left(\frac{\partial w(z,t)}{\partial z}\right)^2}dz = \frac{\rho U^2 l\alpha^2}{2}\pi\frac{s}{l}\mu^2(t)b^2; \tag{50}$$

Longitudinal force without concentrated force

$$X_1 = -\alpha_1\int_{s-l}^{s} p_1 dx + \alpha_2\int_{s-l}^{s} p_2 dx = -\alpha Y - \frac{\rho U l\beta^2}{2}R, \tag{51}$$

where R is the drag of the symmetric wedge

$$\begin{aligned} R &= \frac{2}{\pi l}\int_{s-1}^{s}\ln\frac{bs+\sqrt{x^2-(s-l)^2}}{\sqrt{s^2-x^2}}dx \\ &= \frac{2}{\pi l}[(1+b)\ln(1+b)+(1-b)\ln(1-b)]. \end{aligned}$$

Drag of an asymmetric wedge is equal to $D = X_1 + X_0$. The pressure center is

$$\lambda = s\left(1-\frac{\pi b^2}{4[E(b)-(1-b^2)K(b)]}\right). \tag{52}$$

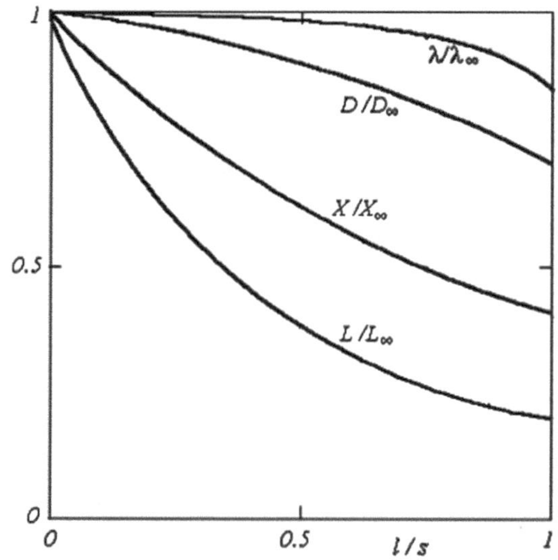

Fig. 7 Dependency of hydrodynamic parameters of wedge on submergence depth

When distance s between the wedge and the free water surface approaches infinity as ($s \to \infty$), all hydrodynamic parameters approach the limits corresponding to the Kirchhoff's model: i.e. lift $L_\infty = \pi\rho V^2 l$, drag $D_\infty = -\rho V^2 \beta^2 l/\pi$, and pressure center $\lambda_\infty = 1/4$. Figure 7 shows dependencies of ratios L/L_∞, D/D_∞ and λ/λ_∞ on length-to-immersion ratio l/s.

7 The Green Functions in the Linear Theory

Analytical functions of complex variable allow to derive analytical solutions on boundary value problems in many cases or to offer effective numerical algorithms. In some cases, analytical methods can be applied to axisymmetric and even to three-dimensional problems though the latter requires additional research. Nevertheless, there are many other analytical and numerical methods which apply to investigation of flow problems. Below, time-depended Green function will be considered in connection with problems of entry in the gravity liquid. For the sake of completeness a brief description is given here to applications of Green function in problems of thin bodies entering water at finite Froude numbers.

The first who considered the effects of finite Froude number during the water entry and exit of thin bodies was Moran [11]. He obtained the calculated data for biconvex flat and axisymmetric bodies crossing the water surface vertically. Later, Yim [5] obtained analytical and numerical results for many problems of water entry both with and without gravity taken into account but for infinitely depth. He used time-depended Green function. Another approach with application of Green function was offered by Porfir'ev [12] who considered a number of problems on gravity effects on vertical movement of thin flat and axisymmetric bodies. Fourier transformation has been used in [13]. Two of these approaches, slightly modified and added with numerical calculation, are discussed below.

Let the boundary of a moving body is given by equations $y = \pm f(x - s(t))$, Fig. 8, which satisfy all required conditions of smallness. Then the entry problem is determined by initial-boundary value problem for the speed potential satisfies the following conditions:

- On the y-axis, $y = \pm\, 0,\, x \in (0, h)$,

$$\varphi_y = \mp\, \dot{s} f'(x - s)[\theta(x) - \theta(x - s)]; \tag{53}$$

- On the bottom,

$$x = h,\ y \in (-\infty, \infty)\ \varphi_x = 0; \tag{54}$$

- On the free boundary, $x = 0,\, y \in (-\infty, 0) \cup (0, \infty)$,

$$\eta_t - \varphi_x = 0, \quad \varphi_t - g\eta = 0, \tag{55}$$

where g is the gravity acceleration, $\eta(y)$ is the free height, $\theta(x)$ is the Heaviside function. Besides, the speed potential and its time-derivative should vanish at the initial time and in infinity at any time,

$$\begin{aligned} &\varphi(x, y, 0) = \varphi_t(x, y, 0) = 0, \\ &\varphi(x, y, t) \to 0,\ y \to \pm\infty. \end{aligned} \tag{56}$$

Using time-dependent Green function $G(\xi,\ \eta,\ \tau | x,\ y,\ t)$ that satisfies Laplace's equation in the flow domain and initial-boundary conditions on the real axis as

$$\begin{aligned} &G_{tt} - gG_x = 0, \quad y = 0; \\ &G = G_t = 0, \quad t = \tau, \quad y = 0, \end{aligned} \tag{57}$$

and behaves like $\ln r$ for $x \to \xi,\ y \to \eta$ with $r = \sqrt{(\xi - x)^2 + (\eta - y)^2}$. The partial derivative of speed potential is found by Green's formula as follows

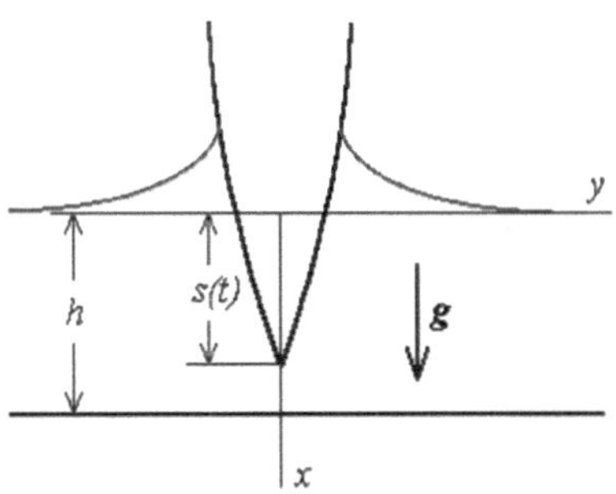

Fig. 8 The sketch of an entry of wedge into gravity water of a finite depth

$$\varphi_t(x, y, t, \tau) = \frac{1}{2\pi}\left(\int\limits_C \varphi_t G_n ds - \int\limits_C \varphi_{tn} G ds\right). \tag{58}$$

Here boundary C coincides with axes y and x that correspond to the free boundary and the body's wetted boundary. Unknown Green function, G, can be obtained as follows:

$$G(\xi,\ \eta,\ \tau|x,\ y,\ t) = \ln\frac{r}{R} + F(\xi,\ \eta,\ \tau|x,\ y,\ t), \tag{59}$$

where $R = \sqrt{(\xi + x)^2 + (\eta - y)^2}$, so that the first term in (59) determines the source potential under the free boundary.

Substituting the latter equation into Eq. 57, one may obtain the partial differential equation on y-axis as

$$B_{tt} - gB_x = -\frac{2g\xi}{\xi^2 + (\eta - y)^2}, \quad x = 0,$$

Using integral identity

$$\frac{\xi}{\xi^2 + (\eta - y)^2} = \mathrm{Re}\int\limits_0^\infty e^{-[\xi + i(\eta - y)]k} dk$$

and setting

$$F = \int\limits_0^\infty T(t, s) e^{-[\xi + x + i(\eta - y)]k} dk, \tag{60}$$

one obtains an ordinary differential equation at the free boundary, $x = 0$,

$$T_{tt} + gkT = -2g \tag{61}$$

with initial conditions $T = T_t = 0$ for $t = \tau$. A solution of Eq. 61 is expressed by function:

$$T(t, k) = -\frac{2}{k}\{1 - \cos[\sqrt{gk}(\tau - t)]\}. \tag{62}$$

Thus,

$$G = \ln\frac{r}{R} - 4\int\limits_0^\infty \frac{e^{-k(\xi + x)}}{k} \sin^2\left(\sqrt{gk}\frac{\tau - t}{2}\right) \cos[k(\eta - y)]\, dk \tag{63}$$

The speed potential obtained in [5] can be found from Eq. 58 as

$$\varphi(x,y,t) = \frac{1}{2\pi}\int_0^t\left(\int_0^\infty\Big([\varphi_\tau]G_\eta - [\varphi_{\tau\eta}]G\Big)d\xi\right)d\tau, \tag{64}$$

where symbol [.] denotes a jump of the value by crossing the x-axis.

The boundary value of partial derivative $[\varphi_{\tau\eta}]$ can be obtained from (53) but partial derivative $[\varphi_\tau]$ is unknown for an arbitrary foil. For a symmetrical wedge, however, since φ_t and G_η are both symmetrical with respect to the x-axis, the first integral of Eq. 64 vanishes, and the speed potential and all dynamic parameters could be calculated completely. From the boundary condition (53), one obtains

$$\begin{aligned}[\varphi_{\tau\eta}] = &-2\ddot{s}f'(x-s) + 2\dot{s}^2 f''(x-s) - \\ &-2\dot{s}^2 f'(x-s)\delta(x-s), \quad x\in(0,\ s), \quad y=\pm 0\,,\end{aligned} \tag{65}$$

where δ is the Dirac delta function. The potential can be calculated by double integrating with respect to $\xi \in (0,\ x)$ for $\eta = 0$and to time $\tau \in (0,\ t)$.

Let a flat-sided wedge with open angle 2α moves at a constant speed V, so that the submergence depth is $s = Vt$. Due to condition (65), potential (64) can be written in the form

$$\varphi = \frac{\alpha}{\pi}V^2\int_0^t G(\tau,0,\tau|\, x,y,t)d\tau. \tag{66}$$

Whence the partial time-derivative is

$$\varphi_t = \frac{\alpha}{\pi}V^2\left(G(t,0,t|x,y,t) + \int_0^t G_t(\tau,0,\tau|x,y,t)d\tau\right).$$

Now, the dynamic pressure on the wedge side can be calculated by $p(x,t) = -\rho\varphi_t$. Integrating with respect to time, one obtains the final expression for pressure

$$p = \rho\frac{\alpha}{\pi}V^2(p_0 + p_1), \tag{67}$$

where

$$p_0 = \ln\frac{Vt+x}{Vt-x};$$

$$p_1 = 2\int_0^\infty \frac{e^{-kx}}{k(k+g)}\begin{pmatrix} ge^{-ktV} + \sqrt{kg}\sin(\sqrt{kg}\,t) - \\ -g\cos(\sqrt{kg}\,t) \end{pmatrix} dk.$$

Results of computations using two last equations are plotted in Figs. 9 and 10.

They show pressure coefficient $C_p = 2p/\alpha\rho\,\dot{s}^2$ at the wedge and a ratio of drags $\bar{D} = D/D_0$ for different Froude numbers, $Fr = \dot{s}^2/gs$. The drag D is calculated by integrating the pressure distribution along the wedge side; the drag $D_0 = \rho s\dot{s}^2\alpha^2 4 \times \ln 2/\pi$ corresponds to weightless fluid. High accuracy of the integral in p_1 can be obtained even when the upper limit is equal to 50 instead of infinity. As seen in Fig. 9, the pressure is not equal to zero at point $x = 0$ due to rising water level at the wedge. The water level can be calculated using Eq. 55; the water level at the wedge

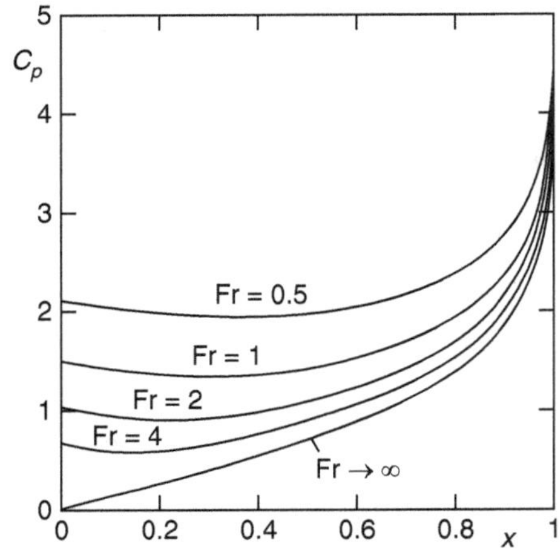

Fig. 9 Pressure distribution at a wedge entering into water

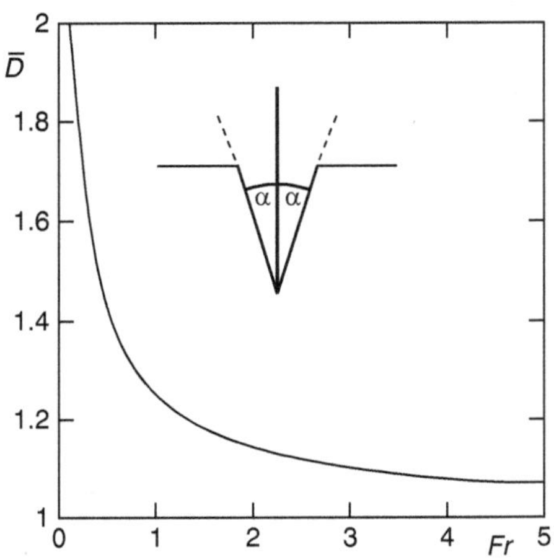

Fig. 10 Drag ratio versus Froude number

is $\eta(0,t) = -p(0,t)/g$. Figure 10 shows that the dynamic drag converges fast to the drag at infinite Froude number.

8 The Fourier Transformation

Setting the speed potential as sum [13]

$$\varphi = \varphi_0 + \varphi_1, \tag{68}$$

where first term is a speed potential of entry into weightless liquid satisfying conditions (53) and (54) and condition $\varphi_0 = 0$ at the free boundary

$$\varphi_0 = \frac{\dot{s}(t)}{2\pi} \int_0^s f'(\xi - s) G(x, y, \xi) d\xi, \tag{69}$$

with

$$G_0(u, y) = \ln \frac{\cosh \frac{\pi y}{2h} + \cos \frac{\pi u}{2h}}{\cosh \frac{\pi y}{2h} - \cos \frac{\pi u}{2h}}.$$

The second term $\varphi_1(x, y, t)$ satisfies Laplace's equation in the flow domain and the boundary conditions as follows:

$$\begin{aligned}
&\varphi_{1y}(x, 0, t) = 0, \quad x \in (0,\ h);\\
&\varphi_{1x}(h, y, t) = 0, \quad y \in (-\infty,\ \infty);\\
&\varphi_{1tt}(0, y, t) - g\varphi_{1x}(0, y, t) = g\varphi_0(0, y, t),\ y \in (-\infty,\ \infty);\\
&\varphi_1 \to 0, \quad y \to \pm\infty;\\
&\varphi_1(0, y, 0) = 0, \quad \varphi_{1t}(0, y, 0) = 0.
\end{aligned}$$

Using Fourier even transformation

$$\varphi_1 = \frac{2}{\pi} \int_0^\infty [A(k, t) \cosh kx + B(k, t) \sinh kx] \cos ky\, dk,$$

and satisfying above mentioned boundary conditions, one can obtain the factor, $B = -A \tanh kh$, the ordinary differential equation as

$$A_{tt} + \omega^2 A = F(t), \quad \omega = \sqrt{kg \tanh kh} \tag{70}$$

with initial conditions $A(k,0) = A_t(k,0) = 0$. The solution of Eq. 70 is

$$A(k,t) = \frac{1}{\omega}\int_0^t F(k,\tau)\sin\omega(t-\tau)\,d\tau, \tag{71}$$

where

$$F(k,t) = \frac{g\dot{s}}{\cosh kh}\int_0^s f'(\xi - s)\cosh k(h-\xi)\,d\xi.$$

Now, the second function in (68) as well as other hydrodynamic parameters can be calculated. Namely, the drag coefficient due to gravity is represented as

$$\begin{aligned} C''_D = {} & \frac{8g}{\pi h}\int_0^\infty \frac{(\cosh[k(h-s)] - \cosh(hk))\cos(\omega s)}{\cosh(hk)} F + \\ & + \frac{8}{\pi h}\int_0^\infty (\sinh[k(h-s)] - \sinh(hk))\omega\sin(\omega s) F dk\ , \end{aligned} \tag{72}$$

where

$$F = \frac{\sinh[k(h-s)] - \sinh kh}{k^2[k\cosh kh + g\sinh kh]}.$$

Using another approach Kuznezov and Manevich [14] obtained results similar to Eq. 72, which can be transformed asymptotically to Yim's results.

The water entry without gravity force is determined by speed potential (69). Coefficients of pressure and drag in this case are calculated by

$$C'_p = \frac{1}{\pi}G(x,0,s), \quad C'_D = \frac{2}{\pi s}\int_0^s G(x,0,s)\,dx. \tag{73}$$

Figure 11 shows the effect of immersion depth on drag of the wedge entering weightless water with the finite depth. Coefficient $C'_D o = C'_D(h \to \infty) = 8\ln 2/\pi$ corresponds to infinite depth; coefficient $C^1_D = C'_D(s = h) = 2.970$ corresponds to the case when the wedge top touches the bottom.

An effect of Froude numbers on drag coefficient C''_D is shown in Fig. 12. It should be noted, that the full drag of the wedge consists of dynamic drag, which is proportional to the angle square, α^2, and hydraulic pressure $p_h = \rho g x$, which is proportional to gravity acceleration g and angle α. Thus, the dynamic pressure and

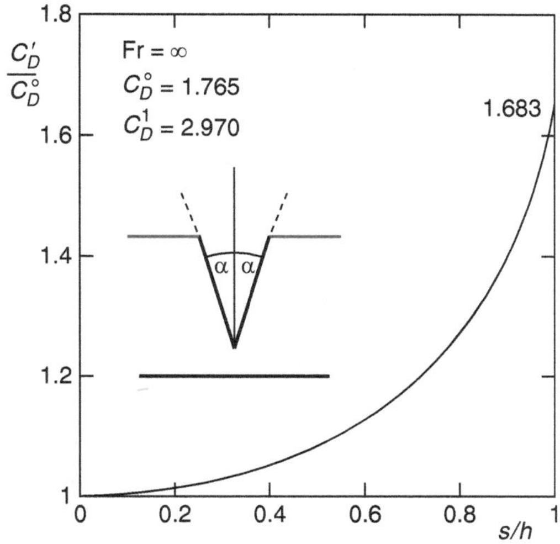

Fig. 11 Effect of depth on drag for a wedge entering a weightless fluid

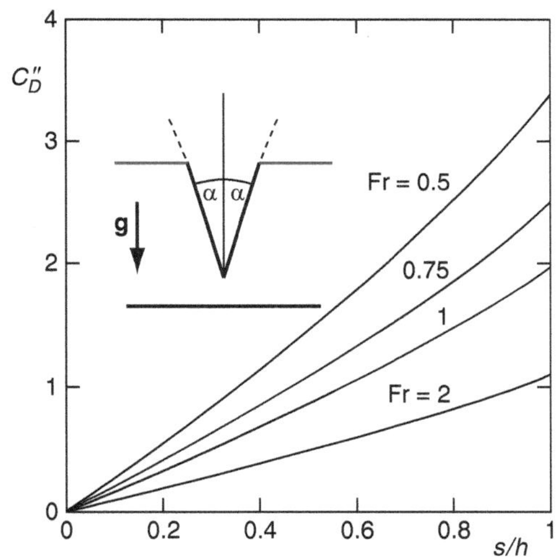

Fig. 12 Gravity effect for a wedge entering finite depth water

dynamic drag should be taken into account if $Fr \approx 1/\alpha$, i.e. Froude number should be quite large. Otherwise it suffices to take into account the hydraulic pressure only.

References

1. Logvinovich GV. Hydrodynamics of streams with free boundaries. Kiev: Nauk. Dumka; 1969 (In Russian).
2. Sagomonyan AYa. Penetrating. Moscow University; 1974 (In Russian).

3. Korobkin AA, Pukhnachov VV. The initial stage of water impact. Ann Rev Fluid Mech. 1988;20:159–95.
4. Terentiev AG. Problems in the theory of high-speed hydrodynamics. Proceedings of the International Summer Scientific School "High Speed Hydrodynamics". Cheboksary; 2002. p. 11–27.
5. Yim B. Investigation of gravity and ventilation effects during the water entry of thin foils. Proceedings of the IUTAM Symposium "Unsteady Flow of Water at High Speed". Leningrad; 1971. p. 6.
6. Terentiev AG. Inclined entry of a thin body into incompressible liquid. Izvestia AN SSSR MZhG. 1977b;5:16–24 (In Russian).
7. Terentiev AG. Inclined entry in ideal non-gravity liquid of thin body with ventilated cavity. Izvestia AN SSSR MZG. 1979;3:66–76 (In Russian).
8. Gradstain IS, Ryzhik IM. Tables of integrals, sums, series and products. M. GIFML; 1962.
9. Dobrovolskaya ZN. On some problems of similarity flow of fluid with a free surface. J Fluid Mech. 1969;36(part 4). pp. 805–829.
10. Terentiev AG, Mikhailov VM. Non-stationary moving of the slender bodies in an ideal fluid. Proceedings of the Non-stationary moving of bodies in a fluid. Cheboksary; 1979. p. 111–48. (In Russian).
11. Moran JP. The vertical water-exit-and-entry of slender symmetric bodies. J Aerospace Sci. 1961;28:803–12.
12. Porfiriev NP. Vertical moving of a thin axisymmetric body to free surface of a gravity liquid. Proceedings of the "High Speed Hydrodynamics". Cheboksary: Chuvash State University; 1981. p. 100–9. (In Russian).
13. Galanin AV, Terentiev AG. Boundary value problems of linear hydrodynamics. Cheboksary: Chuvash University Press; 1984 (In Russian).
14. Kuznetsov AV, Manevich ASh. Gravity effect in water entry of thin foil. Proc. Trudy NKI. Nikolaev: NKI; 1979. p. 152. (In Russian).

Study of the Parameters of a Ventilated Supercavity Closed on a Cylindrical Body

Yu. D. Vlasenko and G. Yu. Savchenko

Abstract

The paper reports the procedure and results of an experimental study of ventilated supercavities on axisymmetric bodies of various geometries in a water tunnel. Hysteresis processes for variously shaped model bodies are considered. The results of recording of unsteady processes of cavity size variation caused by gas injection shutoff are presented. The effect of the body geometry on gas entrainment and unsteady cavity size variation is shown.

1 Introduction

The hydrodynamics of developed cavity flows is widely covered in the literature. The prevailing majority of experimental studies in this field are concerned with regimes of ventilated cavitation, which allow one to study supercavities with small cavitation numbers at relatively low velocities in steady-state conditions. In particular, for these conditions qualitatively different types of gas entrainment have been found [1–3]:

(a) Pulsing mode of gas entrainment with toroidal vortices;
(b) Gas entrainment by vortex filaments;
(c) Continuous gas entrainment with foam.

Special experiments have shown that the third type of entrainment (c) is always concurrent with the first two types, (a) or (b), as a result of which general relations in the process of gas entrainment from cavities are difficult to find [1]. Hysteresis properties characteristic of gas entrainment process in the case of free cavity closure are pointed out in the literature too [4].

Yu.D. Vlasenko (✉)
Institute of Hydromechanics of the National Academy of Sciences of Ukraine, Kyiv, Ukraine
e-mail: Vlasenko34@gmail.com

I. Nesteruk (ed.), *Supercavitation*,
DOI 10.1007/978-3-642-23656-3_11,

At the same time, there is insufficiency of experimental data, on the character of gas entrainment in the case of developed cavities closure on bodies. Such information is of interest in terms of both pure science and applications, and thus for experiments along this line it is advisable to use the simplest and most frequently used cylindrical and conical bodies of revolution.

This paper is concerned to cavities behind the of disc cavitators installed in the fore part of axisymmetric bodies with a cylindrical middle part and conical ends at various body-to-cavitator diameter ratios. The studies conducted involve two questions:

(a) Study of the relationship between the cavity parameters and the gas injection rate for the case of a steady cavity, where the gas injection and gas entrainment rate are in balance. The physical aspect of gas entrainment from a steady cavity is covered in the literature in considerable detail [1–10]. In the present study, the features of evolution of a steady cavity with the variation of the gas injection rate are considered from the standpoint of a possibility to control the cavity parameters.

(b) Study of the time evolution of the parameters of an unsteady cavity caused by an abrupt change in the gas injection rate. This issue is also covered in the literature [11–14]. In the present study, the time evolution of ventilated supercavities on gas injection shutoff is considered. Such processes are of interest from two standpoints. First, the evolution of an unsteady cavity in this case is fully governed by a single factor – the dynamics of gas entrainment from the cavity. Second, during the collapse of a supercavity the maximum possible rate of change of its parameters is achieved, which is also of interest in terms of cavity control.

2 Experimental Procedure

Experimental studies were carried out on the small water tunnel (SWT) at the Hydrodynamic Laboratory, the Institute of Hydromechanics of NASU. A schematic diagram of the test model installed in the SWT working area together with auxiliary equipment is shown in Fig. 1. The central element of the body is a hollow cylindrical rod 1 of diameter 20 mm and length 1.1 m with a cavitator 2 of constant diameter $D_n = 20$ mm installed in its fore part. Replaceable elements were used to install on the rod 1 axisymmetric bodies of various sizes, which included a cylindrical middle part 3 of variable diameter D_b and conical end parts 4 with cone angles of $6° - 8°$. The geometry of the test bodies is shown in Table 1.

A photo of the model with a cavity in the SWT working area is shown in Fig. 2. The flow rate of the air injected into the cavity was measured with a type RS-5 rotameter 5. As shown in Fig. 1, air injection into the cavity was controlled using a valve 6 and electric valve 7, the latter serving to obtain unsteady flow regimes. The electric valve 7 is connected via a control unit 8 to a type ChZ-54 electronic frequency meter 9, which served to time the development of unsteady processes.

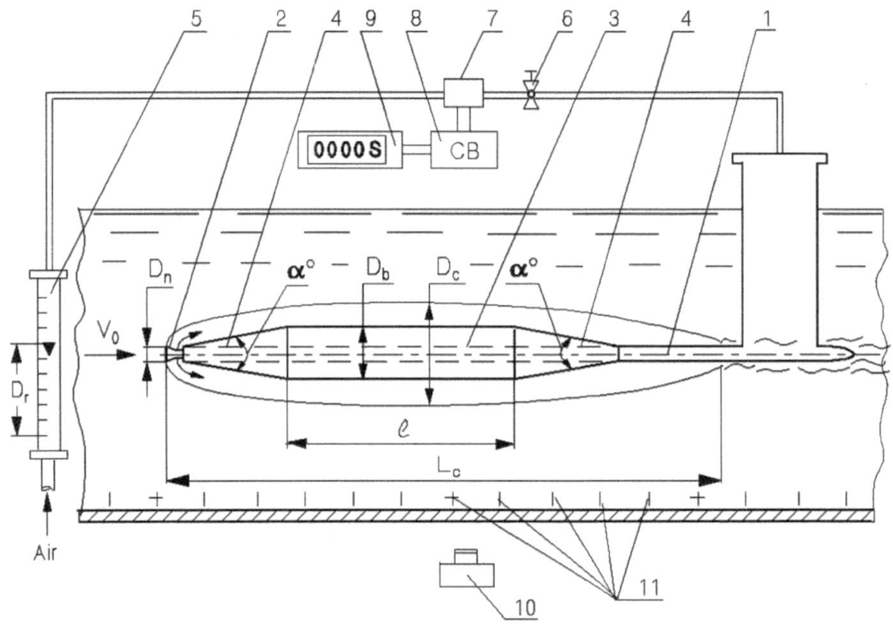

Fig. 1 Schematic of the model and auxiliary equipment in the SWT working area

Table 1 The geometry of the test bodies

D_n, mm	Cylinder D_b, mm	l, mm	Cone α°
20	56	238; 538	7
20	44	685	8
20	32	945	6
20	20	1,100	–

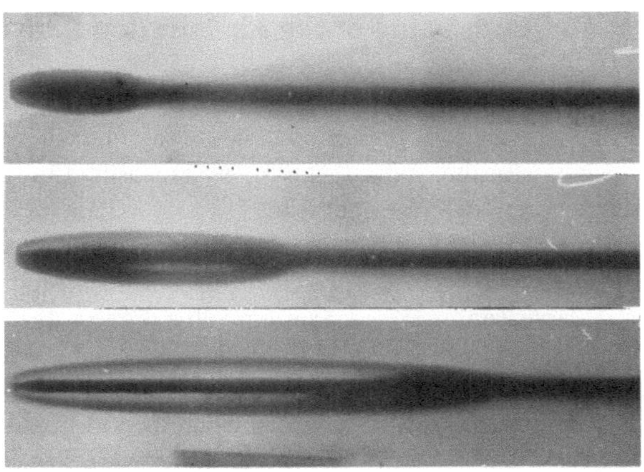

Fig. 2 Photos of a model with a cavity in the SWT working area

Cavity flows were recorded using a SHARP-type video camera 10, model VL-H420, at a shutter speed of 1/50 s. To determine the cavity dimensions from the recorded video data, the glass walls of the SWT working area had a line scale 11 marked every 100 mm.

The experiments were conducted at a constant flow velocity in the SWT working area $V_0 = 8.9\ \mathrm{m/s}$, a constant cavitator immersion depth equal to 180 mm from the free surface, and a constant cavitator diameter $D_n = 20$ mm. At this flow velocity, the effect of gravity is noticeable, which manifests itself in so called cavity surfacing, as a result of which the cavity symmetry axis is at an angle to the flow axis. Because of this, for the measured data to be comparable the test bodies were installed at the corresponding constant angle to the flow axis thus providing identical axisymmetric cavity closure on the body surface in all flow regimes.

In the experiments, the following parameters were changed:

- The geometry of the body downstream of the cavitator according to Table 1;
- The rate of air injection into the cavity over the range $Q_g = 0 - 0.94\ \mathrm{L/s}$
- The air injection regime (stepwise injection regime and air injection shutoff).

Changing the air injection rate changed the cavity geometry and the corresponding cavitation number

$$\sigma = \frac{p_0 - p_c}{\rho\, V_0^2/2}$$

where $V_0 = 8.9\ \mathrm{m/s}$ is flow velocity; ρ is liquid density in the flow; p_0 is pressure in the flow at a depth of 170 mm from the free surface, which corresponds to the mean position of the center of the cavity volume; and $p_c$is the pressure in the cavity. For steady cavities at a fixed air injection rate the cavitation number σ ranged from about 0.038 to about 0.079. As to the air injection regime, the following two types thereof were studied.

2.1 Change of the Parameters of Steady Cavities Caused by a Change in the Air Injection Rate with Flow Stabilization for Each Value of the Air Injection Rate Q_g

In this case, the air injection rate was changed manually in steps $\Delta\, Q_g = 0.08\ \mathrm{L/s}$ using the valve 6 (see Fig. 1). In doing so, for each type of the body downstream of the cavitator the cavity length L_c was measured in two runs. In the first measurement run, the air injection rate was increased from minimum values $Q_g = 0.020 - 0.028\ \mathrm{L/s}$, at which a steady cavity starts forming, to the maximum value $Q_g = 0.94\ \mathrm{L/s}$. Then in the second measurement run the air injection rate was decreased from its maximum value to the value at which the cavity collapses. The steady cavity length L_c for each fixed value of Q_g was measured directly using the line scale 11 provided on the glass walls of the SWT working area. The steady flow regimes were also recorded using the video camera 10. The measured cavity length L_c was plotted versus the air injection rate $Q_g$for each direction of change of Q_g.

2.2 Time Evolution of the Parameters of an Unsteady Cavity Caused by Air Injection Shutoff

In this case, for each body type downstream of the cavitator experiments were run in two stages:

At the first stage, the air injection rate was controlled manually in such a way as to form the largest cavity for the chosen fixed air injection rate. To do so, with the electric valve 7 open (see Fig. 1) the air injection rate was increased to $(Q_g)_{\max}$ using the valve 6, after which the air injection rate was gradually decreased to $Q_g = 0.605\mathrm{L/s}$ and the cavity stabilized at this constant air injection rate in the fixed position of the valve 6. This procedure provided a starting steady cavity that had a size close to the maximum one and at the same time was sensitive enough to changes in the air injection rate.

At the second stage, the electric valve 7 was actuated via the control unit 8 to shut off the air injection into the cavity synchronously with turning on the digital timer/frequency meter 9 to time the evolution of the cavity. The experiment was recorded by the video camera 10 so that the rotameter 5, the digital display of the timer/frequency meter 9 and two line scales 11 on the glass walls of the SWT working area were in picture together with the unsteady cavity under study. This allowed one to measure the cavity length L_c from the recorded video data at known time intervals starting from the air injection shutoff time, which was also seen in picture as an abrupt sinking of the float of the rotameter 5.

The length of the cavity in the process of its collapse was measured every second. Because of perspective distortions in the picture, the value of L_c was measured using two line scales on the walls of the working area, and their average was computed. In view of fluctuations, which are characteristic of ventilated cavities due to the peculiarities of the gas entrainment mechanism [2–4, 9–11] and are all the more probable in the case of an unsteady cavity, each experiment was recorded by the video camera two to three times. For each video, at least two independent measurements of the cavity time evolution were made, and for each type of process the average was computed.

The measured data were processed to give the rate V_L of change of the length L_c of the cavity $V_L = \dfrac{\Delta L_c}{\Delta t}$ for fixed times t from the start of the process, or, what is the same, for fixed values of the length L_c of the unsteady cavity, and the rate V_L of cavity collapse along the body in the flow was plotted versus L_c.

In the plots below, the measured linear quantities are normalized to the cavitator diameter D_n, i.e. $\bar{L}_c = \dfrac{L_c}{D_n}$; $\bar{V}_L = \dfrac{V_L}{D_n}$ (s^{-1}). The air injection rate is shown as the dimensionless air injection coefficient $C_Q = \dfrac{4Qg}{V_o \pi D_n^2}$. For the unsteady regimes studied, the air injection coefficient for the starting steady cavity was $C_Q = 0.216$.

The results of measurement of unsteady cavities were also used in estimating the cavity gas entrainment rate in the characteristic portions of the body profile. The difference of the values of the volume of an unsteady cavity measured one second apart (minus the corresponding part of the body volume on the same interval) characterizes the rate Q_g of gas entrainment from the cavity. In this case, the length of the body portion under consideration is numerically equal to the cavity collapse rate.

3 Experimental Results

3.1 Steady Cavity Flow Regimes

Figures 3 and 4 show experimental dependences of the length of steady ventilated cavities on the air injection rate $\bar{L}_c = f\,(C_Q)$ for bodies of various profiles. By and large the plots are identical in form, and they show that the cavity length is not a single-valued function of the air injection rate. This ambiguity is due to the direction of change of the air injection rate, i.e. to hysteresis in the evolution of ventilated cavities.

Each plot includes two branches, which correspond to the increase and decrease of the air injection rate (shown with arrows), and these branches form a characteristic hysteresis loop.

The ascending branch of the plots $\bar{L}_c\,(C_Q)$ is nearly identical for bodies of various profiles, and it shows that considerable air injection rates are needed for a cavity to start forming. It is significant that on reaching some threshold value of C_Q a cavity develops spontaneously to a size close to the maximum one.

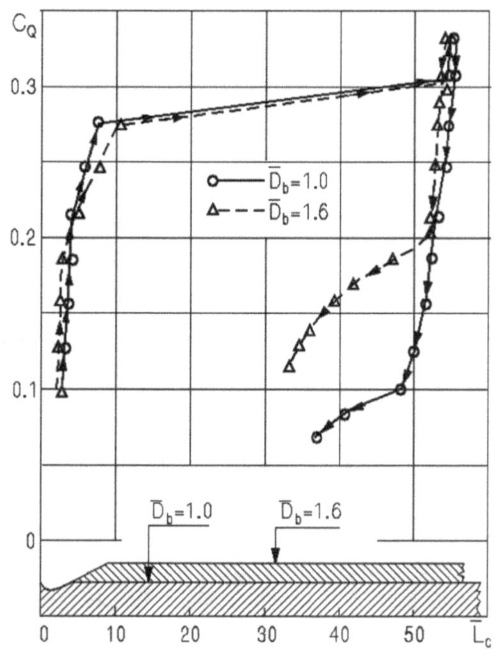

Fig. 3 Plots $C_Q(\bar{L}_c)$ for a steady cavity at $\bar{D}_b = 1.0;\ 1.6$

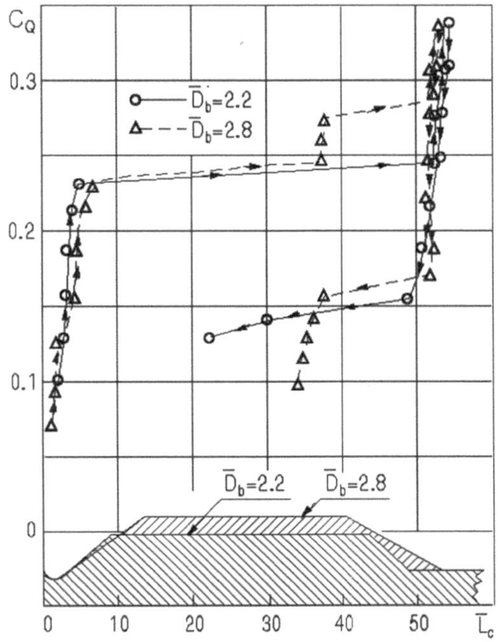

Fig. 4 Plots $C_Q(\bar{L}_c)$ for a steady cavity at $\bar{D}_b = 2.2;\ 2.8$

In its initial portion, the descending branch of the dependence $\bar{L}_c\ (C_Q)$ is also identical in all plots, and it shows that the response of a developed cavity to a sizeable decrease in the air injection rate is insignificant. However, here, too, on decreasing the air injection rate to some threshold value a cavity rapidly diminishes to the point of its collapse. At this stage, the plotted dependences differ quantitatively, which is due to different body geometries in the cavity closure zone.

By and large the plots demonstrate the complexity and limited possibilities of controlling the parameters of a ventilated cavity by varying the air injection rate.

3.2 Unsteady Flow Regimes

Experimental plots of variation of the cavity collapse rate along the body in the flow $\overline{V}_L = f\ (\bar{L}_c)$ are depicted in Figs. 5 and 6, which also show, on comparable scales, the corresponding body profiles. Despite considerable quantitative differences, the plots show that the variation of V_L along the body is identical in form. The process of cavity collapse after air injection shutoff includes three characteristic stages: two maxima of the cavity collapse rate in the zone of the fore and the aft end of the body, respectively, and a minimum of V_L in its cylindrical middle part. A comparative analysis of the plots based on known data on the mechanism of cavity gas entrainment makes it possible to interpret the above features of the process under study as follows.

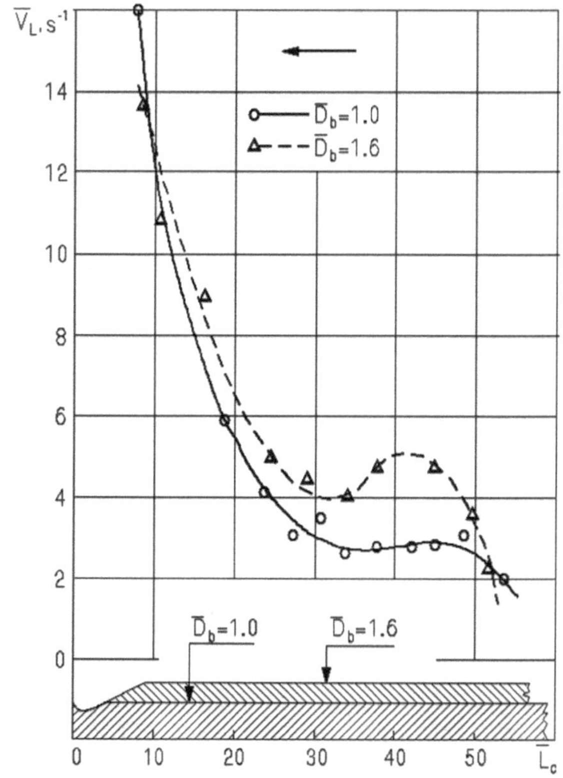

Fig. 5 Cavity collapse rate for $\bar{D}_b = 1.0;\ 1.6$

The first maximum of the cavity collapse rate V_L near the aft end of the body seems to be due to two factors:

- The features of the body geometry, namely, the presence of a conical aft portion (Fig. 6), as a result of which the movement of the wash zone along the cone at the initial stage increases the perimeter around which the gas is entrained from the cavity;
- The features of the process itself because this effect in a weaker form is observed on cylindrical bodies too (Fig. 5). This is confirmed by the data presented below on the variation of the cavity gas entrainment rate Q_g along the body, where the maximum of Q_g is also observed near the aft end of the body (compare with Figs. 9 and 10).

In view of the aforesaid, it seems to be natural to relate the minimum of the cavity collapse rate on the cylindrical middle part of the body to the above factors ceasing to act. As to the abrupt increase in V_L near the fore end of the body, a similar effect was found out in experiments on the collapse of ventilated cavities in free closure conditions [11]. In this case the cavity size also decreases at a progressively increasing rate, and thus it is reasonable to relate this effect mainly to a change of

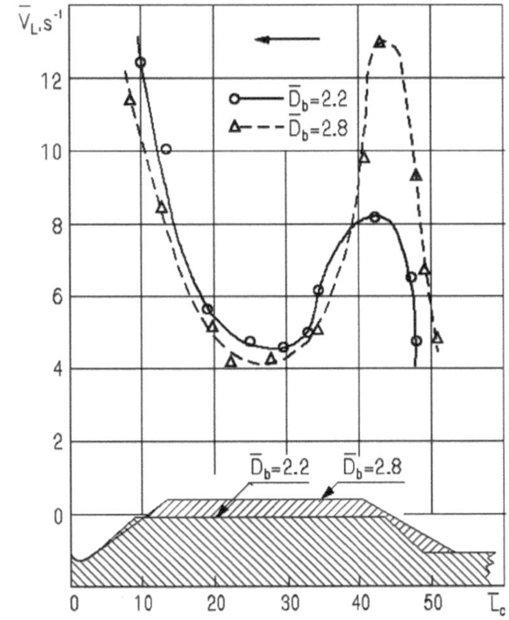

Fig. 6 Cavity collapse rate for $\bar{D}_b = 2.2;\ 2.8$

the cavity gas entrainment mechanism with decreasing relative cavity size. This stage is characterized by a sharp increase in the intensity of re-entrant jets in the cavity closure zone and thus in the level of cavity surface disturbances, as a result of which the gas entrainment rate increases. In this connection, it is pertinent to note that it is technically difficult to determine stable mean values of V_L because the process is fast and the measured parameters undergo sizeable fluctuations due to pulsating gas release from the cavity in the form of foam.

Some generalized idea of the quantitative effect of the geometry of the body in the flow on the cavity closure process can be inferred from the plots in Fig. 7, where the maximum and minimum process rates are plotted versus the body diameter (the maximum values of V_L in the plot relate to the aft zone of the body). It can be seen from the plots that the cavity collapse rate increases with the diameter D_b of the circumference around which the gas is entrained from the cavity. Note that the measured maximum values of the cavity collapse rate are more than an order of magnitude lower than the mainstream velocity V_∞, i.e. $V_L << V_\infty$. This limits the possibility of cavity control by air injection.

At the same time, this order of magnitude of V_L suggests that the gas entrainment process and the time evolution of the cavity size are quasi-steady in character. This allows one to determine the instantaneous volumes of an unsteady cavity as an ellipsoid of revolution [15–17] using the familiar dependences of the geometry of a steady cavity on the cavitation number.

The volume W_c of an ellipsoid of revolution with semi-axes $L_c/2$ and $D_c/2$ is determined as

Fig. 7 Maximum and minimum cavity collapse rate versus model diameter

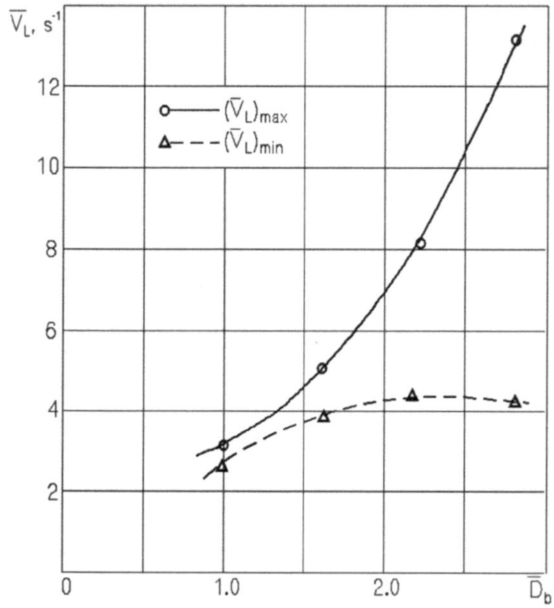

$$W_c = \frac{\pi}{6} L_c D_c^2 = \frac{\pi}{6} \frac{L_c^3}{\lambda_c^2}, \tag{1}$$

where L_c, D_c are the cavity length and maximum diameter, and $\lambda_c = \dfrac{L_c}{D_c}$ is the cavity aspect ratio. One notes that the parameter λ_c of a steady cavity, all other factors being the same, is unambiguously related to the cavitation number. Using the familiar dependence for the diameter of an axisymmetric cavity downstream of a disc

$$\frac{D_c}{D_n} = \sqrt{\frac{C_{xo}(1+\sigma)}{K\sigma}} \approx \sqrt{0.84 \frac{1+\sigma}{\sigma}} \tag{2}$$

and Logvinovich's formula for the cavity length [1]

$$\frac{L_c}{D_n} = \frac{1.92 - 3\,\sigma}{\sigma}, \tag{3}$$

on appropriate substitutions we obtain the semi-empirical dependence $\lambda_c(\bar{L}_c)$

$$\lambda_c = \bar{L}_c \sqrt{\frac{2.3}{\bar{L}_c + 4.92}}. \tag{4}$$

Similar calculations with the use of Epshtein's formula for the cavity length [2]

$$\frac{L_c \sigma}{D_n} = 1.67\sqrt{\frac{C_{xo}}{K}}; \tag{5}$$

give the simpler dependence $\lambda_c(\bar{L}_c)$

$$\lambda_c = 1.53\sqrt{\bar{L}_c}. \tag{6}$$

Here, $C_{xo} = 0.82$ is the disc cavitator drag coefficient at $\sigma = 0$, and $K \approx 0.98$ is an empirical coefficient.

The dependences $\lambda_c(\bar{L}_c)$ obtained by Eqs. 4 and 6 and plotted in Fig. 8 are in rather close agreement with each other and in satisfactory agreement with the data of the experiments described above. The spread of experimental points in the region of small values of $\bar{L}_c$ and λ_c is due to the above-mentioned cavity instability at the final stage of collapse. To determine the instantaneous cavity volume by Eq. 1, the experimental dependence $\lambda_c(\bar{L}_c)$ was used.

The obtained data on the time evolution of gas cavity entrainment are plotted in Figs. 9 and 10 as dependences $C_Q(\bar{L}_c)$; for comparison, the profiles of the corresponding models are shown in the figures too. Here, the plots $C_Q(\bar{L}_c)$ are also qualitatively identical in form, and they show a well-defined maximum in the zone of the aft end of the models. The characteristic feature is that the maximum gas entrainment rate is clearly seen to be located at much the same distance from the cavitator at the section that corresponds to $\bar{L}_c \approx 45$. In a quantitative sense, $(C_Q)_{\max}$ increases with the model diameter. In the zone of the cylindrical middle part of the test bodies, the gas entrainment rate monotonically decreases with a greater or lesser intensity as the collapse progresses.

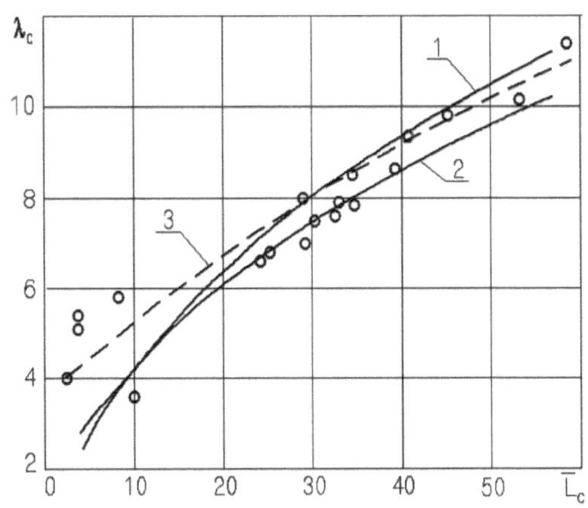

Fig. 8 Graphs of $\lambda_c(\bar{L}_c)$:
1 – Eq. 4; 2 – Eq. 6;
3 – experiment

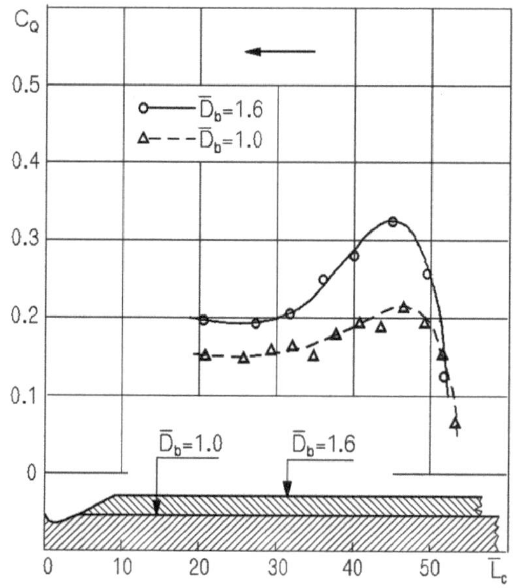

Fig. 9 Time history of cavity gas entrainment at $\overline{D}_b = 1.0;\ 1.6$

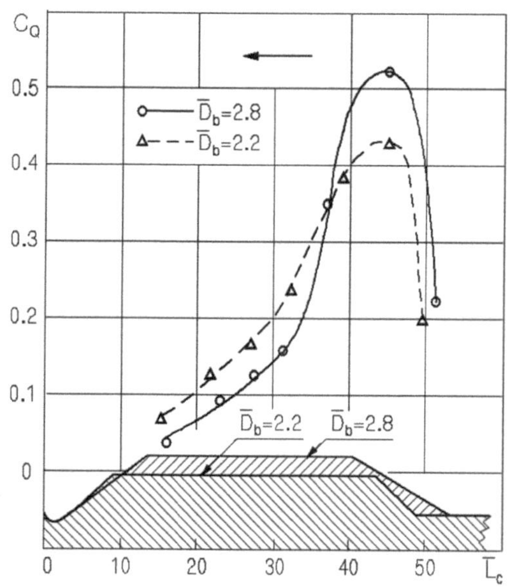

Fig. 10 Time history of cavity gas entrainment at $\bar{D}_b = 2.2;\ 2.8$

The plots $C_Q(\bar{D}_b)$ in Fig. 11 give a more generalized quantitative characteristic of the process under study. Here, a well-defined linear relationship between the maximum gas flow rate and the body diameter (solid line) should be noted. The dependence $C_Q(\overline{D}_b)$ in the fore portion of the model at different distances from the cavitator, but within the cylindrical portion of the body, is shown as dotted

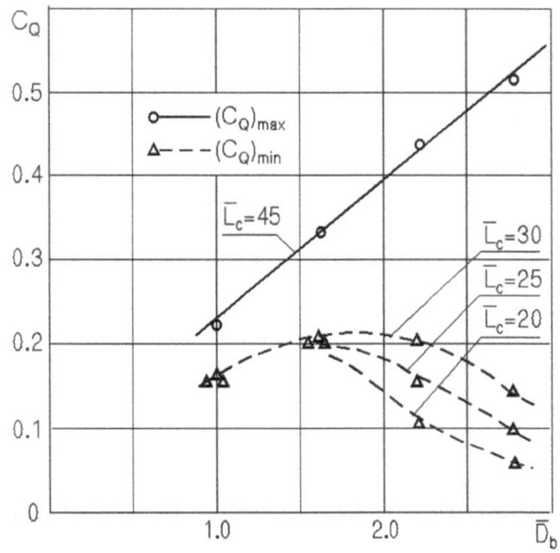

Fig. 11 Plots of $(C_Q)_{\max}$ and $(C_Q)_{\min}$ versus the model diameter

lines. These plots demonstrate the opposite behavior of the dependence $C_Q(\overline{D}_b)$: the gas entrainment rate decreases not only as we approach the cavitator, but also with increasing body diameter. The latter seems to be due both to more intensive gas release early in the cavity collapse and to a relatively smaller initial gas volume in a cavity on large-diameter bodies.

Conclusions

1. Cavity size control by varying the gas injection rate into the cavity is highly problematic because hysteresis is present, the gas injection rate is not a single-valued function of the cavity closure location, and the rate of change of the cavity length governed by gas entrainment is limited.
2. Experiments have shown that the gas injection rate required for sustaining a ventilated cavity closed on a circular cylinder varies monotonically with the cavity length within the cylindrical portion of the model.
3. The evolution of an unsteady cavity on air injection shutoff is characterized by intensive gas release early in the cavity collapse, the maximum gas entrainment rate being a linear function of the body diameter. Later in the cavity collapse, the gas entrainment rate decreases monotonically with a greater or smaller intensity.

References

1. Logvinovich GV. Free-boundary flow hydrodynamics (in Russian). Kiev: Naukova Dumka; 1969.
2. Epshtein LA. Methods of dimensional theory and scaling in vessel hydromechanics problems (in Russian). Leningrad: Sudostroenie; 1970.
3. Epshtein LA. Characteristics of ventilated cavities and some scale effects (in Russian). In: Unsteady high-speed water flows. Moscow: Nauka; 1973. p. 173–85.

4. Michel JM. Ventilated cavities. A contribution to the study of pulsation mechanism. In: Unsteady high-speed water flows. Moscow: Nauka; 1973. p. 343–60.
5. Egorov IT, Sadovnikov YuM, Isaev II, Basin MA. Ventilated cavitation (in Russian). Leningrad: Sudostroenie; 1971.
6. Rozhdestvensky VV. Cavitation (in Russian). Leningrad: Sudostroenie; 1977.
7. Epshtein LA. On the mechanism of pulsation processes in the trailing part of attached cavities (in Russian). Proceedings of the Symposium on the Physics of Acoustic/Hydrodynamic Phenomena. Moscow: Nauka; 1975. p. 133–8.
8. Levkovsky YuL. Cavity flow structure (in Russian). Leningrad: Sudostroenie; 1978.
9. Lapin VM, Epshtein LA. On gas entrainment caused by cavity pulsations (in Russian). Uchenye Zapiski TsAGI. 1984;15(3):23–30.
10. Semenenko VN. Computer simulation of ventilated cavity pulsations (in Russian). Gidromekhanika. 1997;71:110–8.
11. Korolev VI, Vlasenko YuD, Boiko VT. Experimental study of cavity development in unsteady gas cavitation (in Russian). Gidromekhanika. 1973;24:79–93.
12. Savchenko VT, Savchenko YuN. Unsteady motion of a disc in the presence of a collapsing cavity (in Russian). Gidromekhanika. 1976;34:35–8.
13. Karlikov VP, Reznichenko NT, Khomyakov AN, Sholomovich GI. Study of unsteady cavity flows (in Russian). In: Terentiev A.G. (ed.) High-speed hydrodynamics. Cheboksary: Chuvash University; 1985. p. 66–76.
14. Savchenko VT, Savchenko YuN. Study of the time history of the collapse of a ventilated cavity downstream of a disc (in Russian). Gidromekhanika. 1974;30:63–7.
15. Knapp R, Daily J, Hammitt F. Cavitation (in Russian). Moscow: Mir Publishers; 1974.
16. Reichardt H. The laws of cavitation bubbles at axially symmetric bodies in a flow. Ministry of AirCraft Production (Britain); 1946. Rept. and Transl. 766P.
17. Self M, Ripken JF. Steady-state cavity studies in a free-jet water tunnel. St.Anthony Falls Hydr. Lab Rept 47; July, 1955.

Hydrodynamic Performances of 2-D Shock-Free Supercavitating Hydrofoils with a Spoiler on the Trailing Edge

Zaw Win, G.M. Fridman, and D.V. Nikushchenko

Abstract

The objective of this paper is to discuss a new type of a shock-free supercavitating wing with the controllable forward flap at the leading edge and the spoiler mounted on the trailing edge. The viscous-inviscid interactive method is applied to determine the hydrodynamic performances. The controllable forward flap plays a vital role to create cavity covered the whole surface of the upper part at any speed in the operating range. The inviscid flow problem is based on non-linear potential flow theory. The Tulin-Terentiev cavity closure scheme with single-spiral vortex termination is adopted and smooth detachment is taken into account near the stagnation zone in the spoiler vicinity. The parametric study is carried out in the framework of inviscid flow problem. The hydrodynamic performances of the hydrofoil are finalized with RANSE solver of CFD code. Such a type could be used as a section of submerged supercavitating wing.

1 Introduction

Phenomena involved in cavitation are usually highly nonlinear, unsteady, transient, multi-phase, mixing, and phase changing. The study of cavitation near the free surface is primarily within the linear and inviscid scope. The conformal mapping technique is the main solution procedure [1]. Later, the development of lifting-line and lifting-surface theories enables one to extend the study of three-dimensional linearized problems. With the progress of the theoretical development, the nonlinear theories soon dominate the study of the cavitating flow near the free surface.

G.M. Fridman (✉)
State Marine Technical University, St. Petersburg, Russia
e-mail: grifri@peterlink.ru

I. Nesteruk (ed.), *Supercavitation*,
DOI 10.1007/978-3-642-23656-3_12, © Springer-Verlag Berlin Heidelberg 2012

The advance of modern computers brings to rapid development of computational methods. Recently, the rapid development of computational fluid dynamics has made it possible to take into account the effects of viscosity and turbulence. Such progress makes the simulation more realistic. Furthermore, more complicated and practical cavitation models can be incorporated in the approach.

The importance of CFD in cavitation prediction has been increasing. The current multi-phase flow capabilities of some of the more advanced Reynolds Averaged Navier–Strokes (RANS) solvers are being found to be helpful in gaining insights into the cavitation performance viscous-inviscid interactive methods of marine propellers.

In present paper authors introduce a new-type supercavitating hydrofoil. The flow past the shock free supercavitating hydrofoil with stagnation zone in the spoiler vicinity and wedge-like leading edge is based on the theory of jets in an ideal fluid [2–4]. Using the term 'shock free' we assume that the stagnation point coincides with point D (the dividing streamline reaches the wedge-like leading edge in its vertex D) and the upper bound of the cavity smoothly detaches at point B, see Fig. 1.

2 Problem Formulation

Let us consider the nonlinear shock–free cavitating problem for the flat plate with the spoiler [5], at arbitrary cavitation number, see Fig. 2, the influence of gravity being neglected. Tulin-Terentiev cavity closure model with single-vortex termination was adopted. The origin of the Cartesian coordinate system is taken at the

Fig. 1 Sketch of the shock-free supercavitating hydrofoil with spoiler and controllable forward flap at the sharp leading edge

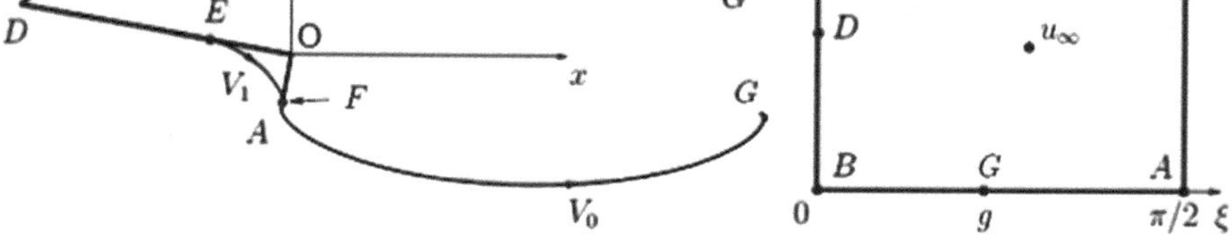

Fig. 2 The physical plane $z = x + iy$ and auxiliary quadrant $\varsigma = \eta + i\xi$

plate's trailing edge, x-axis being directed downstream and y upwards. There is an incident stream with speed V_∞ coming from the left. The region occupied by the fluid is bounded by the rigid boundaries OD, OA and DB and the free surfaces AG, BG, EF, interval DE, FA being unknown length. Note that the interval OD $= l$, OA $= \varepsilon$, and DB $= l_F$. The incidence angle is α, the inclination angle of the spoiler is β and the inclination angle of the forward flap DB is γ. It is of importance that the dividing streamline would meet the cavitating plate at the vertex of the wetted surface BDOA, namely at point D, to provide the shock–free cavitating mode. In this case the length of the leading flap l_F is to be treated as an unknown parameter of the problem, the angle γ being given. Since the cavitation number $\sigma \leqslant 0$, the velocity absolute value on the free surfaces is $V_0 = V_\infty\sqrt{1+\sigma}$. The cavity terminates in spiral vortex placed at points G and G. The velocity V_1 on the boundary of stagnation zone EF is unknown. The zone detaches smoothly (tangentially) at an unknown point E on the plate and smoothly reattaches at another unknown point F on the spoiler. Thus, the problem arising is a two-dimensional (2D) nonlinear problem of the theory of jets in an ideal fluid and its solution can be achieved through the corresponding methods.

With the correspondence between the physical $z = x + iy$ plane and auxiliary quadrant $\varsigma = \eta + i\xi$ shown in Fig. 2, the exact solution can be written out in the form of two derivatives of the complex potential $\omega = \varphi + i\psi$ with respect to the physical and auxiliary variables:

$$\frac{dw}{dz} = V_0 e^{i\alpha}\left(\frac{\theta_1(\varsigma - id_1)}{\theta_1(\varsigma + id_1)}\right)^{\gamma/\pi} e^{\frac{2i\beta\varsigma}{\pi}} \times \exp\left(ic_1\left(\frac{\theta_1(\varsigma)}{\theta_1(\varsigma - g_1)} - \frac{\theta_1(\varsigma)}{\theta_1(\varsigma + g_1)}\right)\right) \tag{1}$$

$$\frac{dw}{d\varsigma} = N_1\theta_1(2\varsigma)\frac{\theta_1(\varsigma - g_1)}{\theta_1^2(\varsigma - \varsigma_\infty)}\frac{\theta_1(\varsigma + g_1)}{\theta_1^2(\varsigma + \varsigma_\infty)} \times \frac{\theta_1(\varsigma - id_1)}{\theta_1^2(\varsigma - \bar{\varsigma}_\infty)}\frac{\theta_1(\varsigma + id_1)}{\theta_1^2(\varsigma + \bar{\varsigma}_\infty)} \tag{2}$$

$$z(\varsigma) = \int_{\frac{\pi\tau}{2}}^{\varsigma}\frac{dz}{d\varsigma}d\varsigma + Z_E = \int_{\frac{\pi\tau}{2}}^{\varsigma}\frac{dz}{dw}\frac{dw}{d\varsigma}d\varsigma + Z_E \tag{3}$$

where $\varsigma_\infty = a + i\,b$ corresponds to the point at infinity at the physical plane (flow domain). Note that a well-known technique of elliptic theta–functions θ_i, $i = 1,2,3,4$, which none $0 < q < 1$ is real ($q = \exp(\pi i\tau)$ $0 < q < 1$), see [4], is proved to be very effective both from the analytical and numerical viewpoints [1, 4]. In the case when the position of the detachment point of stagnation zone Z_E is given the analytical solution includes nine real unknowns a, b, c, d, g, V_1, q, N, and l_F, where $Z_F = l_F e^{-i(\alpha+\beta)}$, which can be determined via the following conditions

$$z(0) = z_B = le^{i(\pi-\alpha)};\ z\left(\frac{\pi}{2}\right) = z_A = \varepsilon e^{-i(\alpha+\beta)};$$
$$\frac{dw}{dz}(u_\infty) = V_\infty;\ \oint\limits_{u_\infty} \frac{dz}{du}du = 0;$$
$$\frac{dw}{dz}\left(\frac{\pi\tau}{2}\right) = V_1 e^{i\alpha};\ z\left(\frac{\pi}{2} + \frac{\pi\tau}{2}\right) = z_A.$$

If position of the point E is unknown then the Brillouin's condition is to be imposed as follows

$$\propto = \frac{d\mu}{ds},$$

where $\mu = \arg(dw/dz)$ denotes the tangential angle to the contour and s is the arc coordinate. It is seen that the Brillouin's condition implies that the stagnation zone detaches smoothly in point E. The flat plate has zero curvature at any point and that is why at point E the second derivative of the function of stagnation zone shape ordinates with respect to x coordinate is to be equal to zero. It is seen that this condition can be readily reduced to

$$\left.\frac{d\mu}{du}\right|_{z_E} = \frac{d}{du}\arg\frac{dw}{dz}\left(\frac{\pi\tau}{2}\right) = 0 \tag{4}$$

3 Results of Non-linear Inviscid Flow Problem

The numerical investigation covered all cavitation number σ less than 1 and the angle of attack α less than 30°. Figure 3 illustrates the lift coefficients, drag coefficients and angle γ versus the cavitation number at any angle of attack α. It is obvious that the hydrofoil enables to create cavity in entire range of cavitation number from smaller to bigger, i.e. not only high velocity inflow but also low one respectively.

Besides it can be seen that hydrofoil provides foil-borne lift for any cavitation number by two means, varying angle γ without changes in angle of attach α and varying the angle of attack α and angle γ simultaneously. The former is more efficient than the latter because lift to drag ratio is considerably high. For instance, such ratios are 9.94 and 1.32 at angles of attack α of 3°and 30° respectively for the cavitation number, $\sigma = 0.21$. As a result, it is noted that the difference in drag coefficients is 16.87 times while lift coefficients just differ in 2.24 times.

Therefore varying only leading edge angle γ is able to achieve highest lift to drag ratio in the whole range of the cavitation number σ, angle of attack α being kept as small as possible. Moreover the hydrofoil can also facilitate a small variant of lift force at any cavitation number through simultaneous adjustment of the angle of

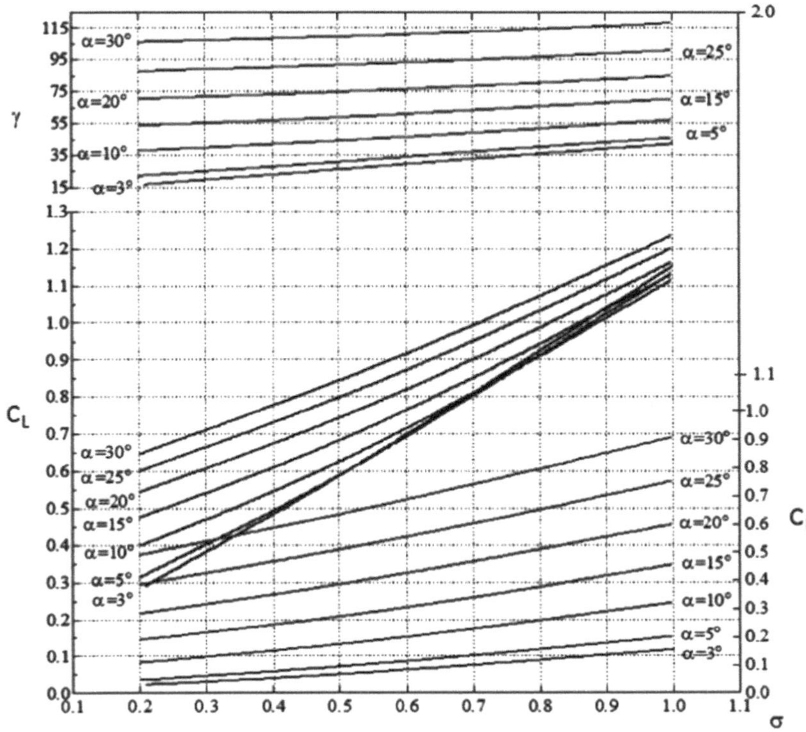

Fig. 3 C_L, C_D and γ versus cavitation number σ for the SC hydrofoil with a parameter set of length $l_F = 0.1$, $\varepsilon/l = 0.02$, $\beta = 90°$

attack α and leading edge angle γ. Meanwhile it should be aware that the hydrofoil thickness affects performance because it has relations with determination of the minimum angle of attack α and the relative spoiler length ratio ε/l.

The thinner the foil thickness is, the better the hydrodynamic fineness will be. On the other hand, there are some restrictions to reduce the foil thickness since the thickness requirement of hydrofoil play vital role in view of strength. Figure 4 illustrates flow patterns for various cavitation numbers σ at a unique angle of attack.

Cavity is shown by grey lines and the boundary of the stagnation zone can be found near spoiler. The problem was solved under the assumption that the Brillouine's condition is satisfied. In figures, the relative height to chord length is about 0.12 between the trailing edge and upper free surface boundary of cavity. The authors left to consider strength viewpoints. However the numerical results reveals that the larger angle α is required for a thicker hydrofoil to keep contact with water vapour only on the entire surface of upper part of foil but it ensures lift to drag ratio will be lower. It should therefore be noted that the larger angle of attack is favoured to obtain enough clearance between upper and lower boundary of cavity for thicker foil thickness while the angle of attack α is required as small as possible so as to

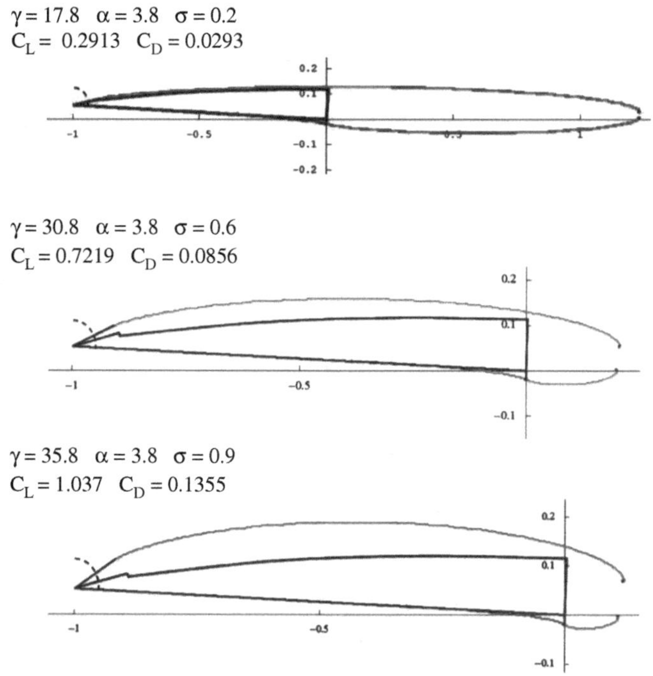

Fig. 4 Flow patterns: shock-free supercavitating hydrofoil with stagnation zone in the spoiler vicinity, and free stream line flow (*gray lines*); relative length of forward flap, $l_F = 0.1$, spoiler length to chord ratio, $\varepsilon/l = 0.02$ and inclination angle $\beta = 908$. Stagnation point coincides with the leading edge. The Brillouine's condition is satisfied

gain optimum the lift to drag ratio. It means that the designer must consider to achieve the best compromise between hydrodynamic and strength aspects.

4 The Viscous Flow Problem with CFD Code

The hydrodynamic performance of resulting hydrofoil from non-linear inviscid flow problem needs to be finalized with CFD method in order to take into account viscous effects. In this paper, the modelling of the viscous flow past the SC foil was carried out by means of RANSE solver of FLUENT when the meshing was complete in GAMBIT. Spalart-Allmaras (S-A) was chosen as a turbulence model. It is more effective than others for two-dimensional multi-phase flow in economic and time consuming viewpoints [6].

First of all, as shown in Fig. 1, it is necessary to configure the body consisting of the lower and upper surface of the foil including spoiler, the forward flap being hinged at the leading edge in GAMBIT. Then the suitable boundary conditions are defined in FLUENT. The mesh elements need to be created about 300,000 including boundary sub-layer nearest to wall with minimum cell thickness, y_P of

0.003 mm. One can determine the minimum cell thickness by using the equation in [7].

$$y_P = y^+ v / \left(u_\infty \sqrt{C_f/2} \right), \text{ where } y^+ < 5 \text{ or } y^+ > 30 \tag{5}$$

Frictional drag coefficient C_f is derived from Reynolds number.

$$C_f/2 = 0.037\mathrm{Re}^{-1/5} \tag{6}$$

The cavitating flow problems are generally solved in two steps, first in single phase flow and second in two phase flows. That is why the cavitating flow problems should be solved after having obtained the reasonable pressure fields around the foil. Single phase flow needs to be converged until the residual of momentum equation is below 10^{-6}. In the first step, the boundary conditions of the flow domain are taken into account with velocity inlet and outflow scheme. After convergent of single phase solution, boundary condition has to be changed with a pair of velocity inlet and pressure outlet and the cavitation mode is turned on. As for second step, the solutions will be reasonable when all of residual are below 10^{-3}. Steady segregated solver is used at single phase flow but unsteady solver is suitable for multi-phase flow.

The results from CFD code reveal that the hydrodynamic characteristics in the frame work of the non-linear inviscid flow seem to be the same as in a viscous flow. The cavity shapes are somewhat different between inviscid and viscous flow due to viscous effect, see Fig. 5.

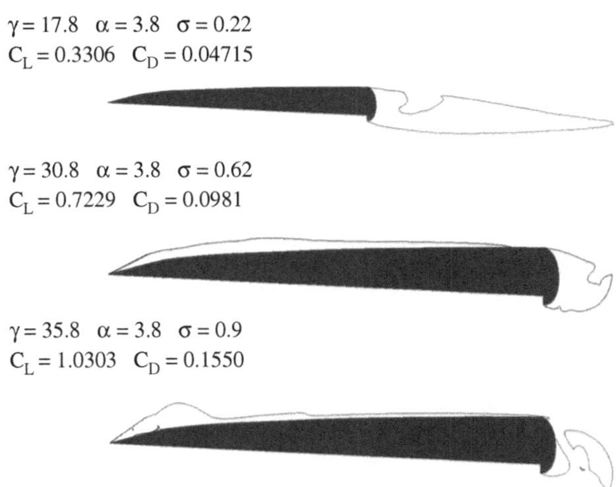

Fig. 5 The cavity around the hydrofoil in viscous flow

Conclusions

The 2-D hydrofoil presented in this paper ensures to achieve supercavitating flow at any speed, low or high. Unfortunately authors left to conduct 3-D flow analysis for this type. Such a hydrofoil can be used as a submerged wing in hydrofoil vessels. Its features are as follows:

(a) Capability of obtaining foil-borne lift even at low speed
(b) Reducing take-off speed
(c) High lift to drag ratio

One can find the 2-D optimum shape of such types of hydrofoil in the frame work of non-linear inviscid flow on the basis of initial parameters such as cavitation numbers in the whole operating range, reasonable relative length of forward flap, possible range of angle γ, minimum foil thickness, inclination angle β, and the relative spoiler ratio. Such a method is not a time consuming one like CFD method but the resulting hydrodynamic characteristics are not too far from that of CFD code.

References

1. Gurevich MI. The theory of jets in an ideal fluid. Moscow: Nauka; 1979 (In Russian).
2. Chaplygin SA. To the problem of jets in a incompressible fluid. Trudy Otdeleniya Phys. Nauk. X(1), Moscow; 1899. (In Russian).
3. Terentiev AG. Mathematical aspects of cavitation. Cheboksary: Chuvash State University; 1981 (In Russian).
4. Achkinadze AS, Fridman GM. Optimal sections for supercavitating propellers with spoiler and preset leading edge angle. St. Petersburg: St. Petersburg State Marine Technical University; 2000.
5. Fridman GM, Uryadov AK. Cavitating flat plate with stagnation zone in the spoiler vicinity. Proceedings of The Second International Summer Scientific School "High Speed Hydrodynamics"; June 2004. Cheboksary; 2004. p. 83–90.
6. Zaw W, Fridman GM. The study of viscous effect on spoiler mounted on the trailing edge of the supercavitating hydrofoil. SubSeaTech 2009; June 2009. St. Petersburg; 2009.
7. Nikushchenko DV. The study of viscous flow of incompressible fluid from the basis of FLUENT, Lecture note (in Russian). St. Petersburg: St. Petersburg State Marine Technical University; 2006.

Index

I. Nesteruk (ed.), *Supercavitation*,
DOI 10.1007/978-3-642-23656-3, © Springer-Verlag Berlin Heidelberg 2012

Kolibri 360 GmbH
Ferdinand-Jühlke-Straße 7
99095 Erfurt, Deutschland
produktsicherheit@kolibri360.de